STUDENT SOLUTIONS MANUAL
to accompany

Chemistry
FOR SCIENTISTS AND ENGINEERS
PRELIMINARY EDITION

Fine • Beall • Stuehr

John Stuehr
Case Western Reserve University

WITHDRAWN

Saunders College Publishing
A Division of Harcourt College Publishers

Fort Worth   Philadelphia   San Diego   New York   Orlando   Austin
San Antonio   Toronto   Montreal   London   Sydney   Tokyo

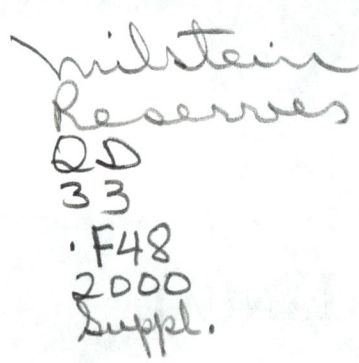
Milstein
Reserves
QD
33
.F48
2000
Suppl.

Copyright © 2000, 1990 by Harcourt, Inc.

All rights reserved. No part of this publication may be reproduced or transmitted in any form or by any means, electronic or mechanical, including photocopy, recording, or any information storage and retrieval system, without permission in writing from the publisher.

Requests for permissions to make copies of any part of the work should be mailed to the following address: Permissions Department, Harcourt, Inc., 6277 Sea Harbor Drive, Orlando, Florida 32887-6777.

Printed in the United States of America

ISBN 0-03-026018-3

012 202 7654321

Preface

Problem solving is a major component in learning chemistry. The more problems you solve and understand, the better your chances of doing well in the course. To this end, we are providing these detailed worked-out solutions to virtually all the odd-numbered problems in the text. We have not included solutions to the problem categories "Estimates and Approximations" and "Writing about Chemistry" since doing so would largely remove the point of the problems, which is to encourage independent thinking about chemistry and its applications. This manual provides not only the worked-out answers, but insights as to the strategy used to solve the problems. You are encouraged to work more problems than your instructor may assign.

Note about significant figures: You should be aware that an electronic calculator is very much a mixed blessing. On the one hand, it provides easy answers to numerical calculations. On the other hand, it typically gives far more digits than are warranted by the data in a problem. As a consequence, you will typically have to round the calculator answer to the correct number of significant figures. In general, I have followed the guidelines for significant figures and rounding that are presented in Chapter One of the text. I have typically carried extra digits in the intermediate parts of the solutions, and then have properly rounded the final answer. If you think a particular step in a calculation has too many significant figures, this is probably the reason.

Although every attempt has been made to provide accurate answers, it is inevitable that some errors have slipped by. If you find such errors, I would very much appreciate being notified of this. You can do so by writing to me or by sending an e-mail message to jes6@po.cwru.edu.

I wish to thank the several persons who assisted in the preparation of this manual. Ms. Dorothy Swain wrote the solutions for a previous version of this book, and I have freely adapted her work where appropriate. Every solution was checked for accuracy by David Sopchak, whose diligence and patience are gratefully acknowledged. I wish to thank Alisa Lewis for making the drawings, and Alisa Lewis and Mary Kate Haney for their fine typing efforts. And finally, I acknowledge the encouragement and patience of my editors, Kent Porter Hamann and Peter McGahey, during the preparation of this manual.

Best wishes.

John Stuehr

CONTENTS

1.	Matter and Measurement	1
2.	Elements and Compounds	5
3.	Stoichiometry	8
4.	Gases	17
5.	Atomic Structure	27
6.	Properties of Molecules	34
7.	Theories of Chemical Bonding	38
8.	Periodic Properties	45
9.	Liquids and Solutions	49
10.	Chemical Equilibrium	60
11.	Acids and Bases	70
12.	Heat, Work, and Energy	85
13.	Spontaneous Change	95
14.	Electrochemistry	102
15.	Chemical Kinetics	113
16.	Solids	118
17.	Materials	122
18.	Properties of Polymers	127
19.	Transition Metals	130
20.	Metallurgy	133
21.	Organic Chemistry	136
22.	Nuclear Chemistry	144

CHAPTER ONE

Matter and Measurement

1.1 When one digit is rounded off, the four numbers become 20., 2.9×10^2, 0.34 and 2.0 respectively.

1.3 One year is 365.25 days long and contains (365.25 days)(24 hr/day)(3600 s/hr) = 3.156×10^7 s.
Therefore, the distance light travels in one year is $(3.156 \times 10^7 \text{ s})(3.00 \times 10^8 \text{ m/s})$ = 9.47×10^{15} m = **9.47×10^{12} km.**

1.5 The numbers in exponential notation are:
a. 2.0500×10^3
b. 2.050×10^{-5}
c. 1.0×10^0
d. 7.27×10^{-1}
e. 6.35×10^1

1.7 0.000000000300   625   0.0625
10,000   0.0001

1.9 a. four   b. three   c. three

1.11 500 sheets corresponds to 50. mm. Therefore, the thickness of one sheet is 50. mm/500 sheets = **0.10 mm/sheet.**

1.13 Because of the number 3×10^{10}, the answer must be reported with **one** significant figure: **6×10^{-4}**

1.15 Use the conversions K = °C + 273.15 and °F = 9/5 °C + 32:
°C = 630.K − 273.15 = **357°C**; °F = 9/5(357°C) + 32 = **675°F**

1.17 We must use a series of unit conversion factors as follows:
a. Convert 100.0 mi/hr to km/hr:
(100.0 mi/hr)(1760 yd/mi)(36 in/yd)(2.54 cm/in)(10^{-3} km/100 cm) = **160.9 km/hr**
b. (100.0 yd)(36 in/yd)(2.54 cm/in)(1 m/100 cm) = **91.4 m**
c. (440 m)(100 cm/m)(1 in/2.54 cm)(1 yd/36 in) = **481 yd**
d. (10,000.0 km)(1000 m/km) = **1.00000×10^7 m**
e. 186,000.0 mi/s)(1760 yd/mi)(36 in/yd)(2.54 cm/in) = **2.993×10^{10} cm/s**

CHAPTER ONE

1.19   a. (100.0 yd)(36 in/yd)(2.54 cm/in)(1 m/100 cm) = 91.4 m which is less than 100 m.
So **100.0 yd is less than 100.0 m**.
b. (400.0 yd)(36 in/yd)(2.54 cm/in)(1 m/100 cm) = 366 m which is less than 400.0 m.
So **it will take longer for the same swimmer to swim 400.0 m.**

1.21   If one metric ton (1000. kg) of coal is 5% sulfur, then it contains (0.05)(1000. kg) = 5×10^1 kg = **5×10^4 g sulfur**.

1.23   Density is defined as d = m/V.
First, convert the volume to milliliters:
V = (one liter)(1×10^3 mL/1 L) = **1×10^3 mL** (exactly).
Next, calculate the mass using m = dV:
m = (0.79 g/mL)(1×10^3 mL) = **7.9×10^2 g** at 20°C

1.25   In order to calculate the total mass of a flask and the liquid that it contains, we must add the mass of the liquid (m = dV) to the mass of the flask (31.3 g). The densities of water, alcohol and mercury can be found in Table 1.7.
a. [(0.997 g/mL)(25.00 mL) + 31.3 g] = **56.2 g** (water + flask)
b. [(0.789 g/mL)(25.00 mL) + 31.3 g] = **51.0 g** (alcohol + flask)
c. [(13.6 g/mL)(25.00 mL) + 31.3 g] = **371 g** (mercury + flask)

1.27   Density = m/V = (1000. g)/(51.8 mL) = **19.3 g/mL**. From Table 1.7, we see that this is the density of pure gold (Au).

1.29   We need to convert units of ($/gallon) to units of ($/kg):
($18/5 gal)(1 gal/3.785 L)(1 L/10^3 mL)(1 mL/0.792 g)(10^3 g/kg)($18/5 gal) = **$1.20/kg**

1.31   Convert $/oz to $/g:
($440.26/oz)(16 oz/454 g) = **$15.52/g**

1.33   The fixed reference points are
        0°C = 32°F
        100°C = 212°F
From these we see that a temperature change of 100°C corresponds to a temperature change of 180°F. Therefore, the ratio of degrees is 180°F/100°C = 9/5. The two scales are linearly related:
    °F = (9/5)°C + constant
When the centigrade reading is 0°C, the Fahrenheit reading is 32; therefore, the constant is 32 and the full relationship is
    °F = (9/5)°C + 32

CHAPTER ONE

1.35 We know that flapability is proportional to the number of flops times potency. The proportionality constant is 0.00849 flips/flop·pot.
Therefore, flapability = (0.00849 flips/flop·pot)(8.47 flops)(203 pots)
Flapability = 14.6 flips

1.37 In order to convert density as g/cm^3 to kg/m^3, we need to use the following conversions:
10^3 g = 1 kg and (100 cm)^3 = (1 m)^3 = 10^6 cm^3 = 1 m^3. Therefore,
(1.625 g/cm^3)(1 kg/10^3 g)(10^6 cm^3/1 m^3) = **1.625 × 10^3 kg/m^3**
We want to convert gallons to pounds for perchloroethylene:
(5 gal)(4 qt/gal)(0.946 L/1 qt)(10^3 mL/1 L))(1.625 g/1 mL)(1 lb/454 g) = **70 lb**

1.39 We are given the following reference temperatures:
0°D = –103°C = 9/5(–103°C) + 32 = –153.4°F
100°D = –35°C = 9/5(–35°C) + 32 = –31.0°F
A change in temperature of 100 degrees D equals –35°C – (–103°C) = 68°C
or (–31°F – (–153.4°F)) = 122.4°F.
∴ 1°D temp change = 0.68°C or 1.224°F and 0°D = –103°C = –153.4°F.

Therefore, the temperature scales are related as follows:
°D = (°C + 103)/0.68   °D = (°F + 153.4)/1.224
Absolute zero is –273.15°C, which is equivalent to °D = (–273.15°C + 103)/0.68
= **–250.22°D**

1.41 1.5 × 10^2 g is the mass of water displaced by the crown.
The volume of the crown is equal to the volume of water it displaces. We can calculate this from the mass of water displaced by using V = m/d.
V_{crown} = V_{water} = m_{water}/d_{water} = (1.5 × 10^2 g)/(1.00 g/mL)
= 1.5 × 10^2 mL = 1.5 × 10^2 cm^3.
Now we can calculate the density of the crown by using d = m/V:
d_{crown} = (2.80 × 10^3 g)/(1.5 × 10^2 cm^3) = **19 g/cm^3**
According to Table 1.7, the density of pure gold is 19.3 g/mL. Apparently the crownmaker was truthful. The measured density of the crown is equal to the density of gold to the two significant figures allowed by the precision of the experiment.

1.43 Phenol is produced at a rate of 400. million pounds per year, which corresponds to
(400. × 10^6 lb/yr)(1 yr/365.25 days) = **1.10 × 10^6 lb/day**
Each oxidizer produces one fourth of this, or 2.74 × 10^5 lb/day per oxidizer. The total hourly rate of production for five oxidizers is
(2.74 × 10^5 lb/day·oxidizer)(5 oxidizers)(0.454 kg/1 lb)(1 day/24 hr)
= **2.59 × 10^4 kg/hr**

CHAPTER ONE

1.47 We see that every increment of 14 units in Q corresponds to an increment of 56 units in R. Therefore, the relationship is linear and will fit the formula R = (56/14)Q + b, where b = a constant. Since R = 842 when Q = 0, b must equal 842. So the relationship is **R= 4Q + 842**.

1.49 a. The densities of copper and zinc are 8.92 g/cm^3 and 7.14 g/cm^3 (Table 1.7). If we assume the alloy is an intimate mixture of Cu and Zn, then its density will be the fraction of each metal times its density. Let f be the fraction of Cu in the alloy. Then
  8.00 g/cm^3 = (f)8.92 g/cm^3 + (1 – f)7.14 g/cm^3
  0.86 g/cm^3 = 1.78(f) g/cm^3
  f = 0.48
The alloy is **48% Cu, 52% Zn**.
b. If the actual density of the alloy is higher than for an intimate mixture, the calculated fraction of copper in the alloy will be too high since the solid solution would require a lower fraction of copper than calculated to achieve the same density.

CHAPTER TWO

Elements and Compounds

2.1 All these atoms are electrically neutral, so the number of electrons equals the number of protons. The number of neutrons is given by $N = A - Z$.
  a. $^{19}_{9}F$ has **9 electrons, 9 protons, and 10 neutrons**.
  b. **20 electrons, 20 protons, and 25 neutrons**
  c. **19 electrons, 19 protons, and 20 neutrons**
  d. **13 electrons, 13 protons, and 14 neutrons**

2.3 a. $Z = 238 - 146 = 92$; element is ^{238}U b. $Z = 22 - 13 = 9$; element is ^{22}F
  c. $Z = 26$; element 26 is Fe   d. $Z = 43$; element 43 is Tc

2.5 a. $^{51}_{28}Ni$ b. $^{29}_{12}Mg$ c. $^{71}_{35}Br$ d. $^{87}_{37}Rb$

2.7 Follow the naming rules for covalent compounds.
  a. sulfur dioxide   b. silicon tetrachloride   c. iodine monochloride

2.9 a. boron tribromide   b. carbon disulfide   c. phosphorus pentafluoride

2.11. These are all binary covalent compounds, the formulas follow directly from their names.
  a. CO_2 b. BrF_3 c. CCl_4

2.13 These are all binary covalent compounds, the formulas follow directly from their names
  a. ICl   b. XeF_4 c. CH_4

2.15 a. Group IA elements form ions by losing one electron: Na^+
  b. Group VA elements form ions by gaining three electrons: P^{3-}
  c. Group VIIA elements form ions by gaining one electron: Br^-
  d. Group IIA elements form ions by losing two electrons: Sr^{2+}

2.17 Formulas of ionic compounds are determined by the charges on the ions.
  a. The ions are Na^+ and S^{2-}; the formula is Na_2S
  b. The ions are Ba^{2+} and O^{2-}; the formula is BaO
  c. The ions are Mg^{2+} and O^{2-}; the formula is MgO

2.19 a. magnesium iodide, MgI_2 b. potassium fluoride, KF
  c. barium chloride, $BaCl_2$

CHAPTER TWO

2.21 Follow the rules for naming ionic compounds; prefixes are not used:
    a. MgO is magnesium oxide   b. LiBr is lithium bromide
    c. RaBr$_2$ is radium bromide   d. Cs$_2$O is cesium oxide

2.23 Names of these ionic compounds follow directly from their formulas; you do need to know the names of the various oxyanions.
    a. potassium sulfate   b. magnesium hydroxide
    c. barium carbonate   d. lithium nitrate

2.25   a. Mg(NO$_2$)$_2$ b. CaCO$_3$ c. K$_2$SO$_4$ d. NaClO$_3$

2.27 Molecular or formula mass is the sum of the atomic masses. Use the table of atomic masses of the elements on the inside cover of the text.
    a. For the P$_4$ molecule, the molecular mass is 4(30.9738) = 123.90 amu.
    b. For C$_2$H$_4$O$_2$, the molecular mass is 2(12.01) + 4(1.008) + 2(16.00) = 60.05 amu.
    c. 267.7 amu   d. 811.1 amu

2.29 In order to calculate the average atomic mass of an element, we multiply the fraction of each isotope in nature by that isotope's mass. Then we sum the resulting products:
0.013(203.973) + 0.273(205.974) + 0.200(206.976) + 0.514(207.977)
= average atomic mass of **Pb = 207 amu**

2.31 In order to calculate the average atomic mass of an element, we multiply the fraction of each isotope in nature by that isotope's mass. Then we sum the resulting products:
0.0582(53.940 amu) + 0.9166(55.935 amu) + 0.0219(56.936 amu) + 0.0033(57.933 amu)
= Average atomic mass of **Fe = 55.85 amu**.

2.33 Since the atomic mass of Br based on natural abundance is 79.904 amu and there are two naturally occurring isotopes, the percent composition can be calculated.
    Let f = fraction of Br with mass 78.918 amu and 1 – f = fraction with 80.916 amu. Then
    79.904 amu = f (78.918 amu) + (1 – f)80.916 amu
    79.904 = 78.918 f + 80.916 – 80.916 f
    1.012 = 1.998 f; f = 0.5065 or 50.65%
% composition: **50.65% ^{79}Br and 49.35% ^{81}Br**

CHAPTER TWO

2.35 $A = N + Z$, and all these atoms are electrically neutral, so the number of electrons equals the number of protons.
  a. $A = 70$, $Z = 31$, $N = A - Z = 39$
     protons = electrons = 31, neutrons = 39
  b. $A = 104$, $Z = 46$, $N = A - Z = 58$
     protons = electrons = 46, neutrons = 58
  c. $A = 242$, $Z = 94$, $N = A - Z = 148$
     protons = electrons = 94, neutrons = 148
  d. $A = 40$, $Z = 18$, $N = A - Z = 22$
     protons = electrons = 18, neutrons = 22

2.37 Name as binary covalent compounds:
  a. dinitrogen pentoxide   b. disilicon hexachloride

2.39 a. Potassium will lose one electron; therefore, the ion will be K^+.
  b. Tellurium will gain two electrons; therefore, the ion will be Te^{2-}.
  c. Bromine will gain one electron; therefore, the ion will be Br^-.
  d. Magnesium will lose 2 electrons; therefore, the ion will be Mg^{2+}.

2.41 a. For C_2H_4, $M = 2(12.011 \text{ amu}) + 4(1.0079 \text{ amu})$
     M = 16.043 amu
  b. For $Sr(NO_3)_2 \cdot 4H_2O$, $M = 87.62 + 2(14.0067) + 3(15.9994) + 4[2(1.0079) + 15.9994]$
     M = 283.69 amu
  c. For $B_{10}H_{14}$, $M = 10(10.811 \text{ amu}) + 14(1.0079 \text{ amu})$
     M = 122.221 amu

2.43 a. For SrF_2, $M = 87.62 \text{ amu} + 2(18.9984 \text{ amu})$
     M = 125.62 amu
  b. For LiCl, $M = 6.941 \text{ amu} + 35.4527 \text{ amu}$
     M = 42.394 amu
  c. For CO_2, $M = 12.011 \text{ amu} + 2(15.9994 \text{ amu})$
     M = 44.010 amu

CHAPTER THREE

Stoichiometry

3.1 The number of molecules can be calculated using the mass of the snowball, the molar mass of water, and Avogadro's number:
(180. g)(1 mol/18.02 g)(6.022 × 10^{23} molecules/1 mol)
= **6.02 × 10^{24} molecules**

3.3 For each, convert the molar mass to number of particles by using Avogadro's number.
a. (63.54 g/mol Cu)(1 mol/6.022 × 10^{23} atoms) = **1.055 × 10^{-22} g**
b. (63.13 g/mol B$_5$H$_9$)(1 mol/6.022 × 10^{23} molecules) = **1.048 × 10^{-22} g**
c. (169.90 g/mol SiCl$_4$)(1 mol/6.022 × 10^{23} molecules)(1 × 10^6 molecules)
= **2.821 × 10^{-16} g**

3.5 a. We can determine the number of moles of NaH$_2$PO$_4$ from its mass and formula mass:
(120. g NaH$_2$PO$_4$)(1 mol/119.98 g) = **1.00 mol compound**
b. Each NaH$_2$PO$_4$ contains one Na. Thus, 1.00 mol NaH$_2$PO$_4$ contains **1.00 mol Na**
Similarly, each unit of NaH$_2$PO$_4$ contains:
2.00 mol H, 1.00 mol P, and 4.00 mol O.
c. (1.00 mol Na)(23.0 g/mol) =   **23.0 g Na**
(2.00 mol H)(1.01 g/mol)   = **2.02 g H**
(1.00 mol P)(31.0 g/mol)   = **31.0 g P**
(4.00 mol O)(16.0 g/mol)   = **64.0 g O**
d. (1.00 mol Na)(6.022 × 10^{23} atoms/mol) = **6.02 × 10^{23} atoms Na**
Similarly, there are:
1.20 × 10^{24} atoms H, 6.02 × 10^{23} atoms P, 2.41 × 10^{24} atoms O

3.7 First, we calculate the molar masses of CO, C$_2$H$_4$ and N$_2$; they are all the same:
$M_{CO} = M_{C_2H_4} = M_{N_2}$ = 28 g/mol
Since we have 28 g of each gas, there are 3.0 mol total in the container.
(3.0 mol)(6.022 × 10^{23} molecules/1 mol)
= **1.8 × 10^{24} molecules**

3.9 The volume of the cube is V = l^3 = (10.0 cm)^3 = 1.00 × 10^3 cm^3. Density d = m/V, so
m = d·V = (7.86 g/cm^3)(1.00 × 10^3 cm^3) = 7.86 × 10^3 g = N × (mass per atom).
N = m/mass per atom = 7.86 × 10^3 g/9.3 × 10^{-23} g
= **8.4 × 10^{25} atoms**

CHAPTER THREE

3.11 Every 100.0 g of the compound contains 53.1 g C and 46.9 g O. We can use these masses to determine the mole ratio of C:O in the compound:
(53.1 g C)(1 mol/12.01 g) = 4.42 mol C
(46.9 g O)(1 mol/16.00 g) = 2.93 mol O
The mole ratio of C:O = (4.42 mol/2.93 mol) = 1.51.
We assume that the last digit contains some uncertainty and that the ratio is 3:2 within experimental error. The empirical formula is **C_3O_2**.

3.13 We have the following information:
Sample mass = 2.200 g; mass of Ca = 0.88 g; mass of C = 0.264 g
Therefore, the mass of O = 2.200 g – 0.88 g – 0.264 g = 1.056 g
moles of Ca = 0.88 g/40.08 g/mol = 0.022 mol
moles of C = 0.264 g/12.01 g/mol = 0.022 mol
moles of O = 1.056 g/16.00 g/mol = 0.066 mol
The elements are in the ratio 1:1:3, so the formula is **$CaCO_3$**

3.15 Assume a sample of exactly 100 g for each oxide. Then:
a. (47.99 g O)(1 mol/16.00 g) = 2.999 mol O
(52.01 g Cr)(1 mol/52.00 g) = 1.000 mol Cr
The empirical formula is **CrO_3**.
b. (31.57 g O)(1 mol/16.00 g) = 1.973 mol O
(68.43 g Cr)(1 mol/52.00 g) = 1.316 mol Cr
O:Cr = 1.499 which is 3:2 with experimental error
The empirical formula is **Cr_2O_3**.
c. (23.52 g O)(1 mol/16.00 g) = 1.470 mol O
(76.48 g Cr)(1 mol/52.00 g) = 1.471 mol Cr.
The empirical formula is **CrO**.

3.17 a. Every 100.0 g of compound contains 30.0 g Fe, 64.5 g C and 5.5 g H.
(30.0 g Fe)(1 mol/55.85 g) = 0.537 mol Fe
(64.5 g C)(1 mol/12.01 g) = 5.37 mol C
(5.5 g H/1 mol/1.01 g) = 5.4 mol H
The empirical formula is **$FeC_{10}H_{10}$**
b. The empirical formula mass of $FeC_{10}H_{10}$ is 186 g/mol. Since this is also equal to the true molar mass, **$FeC_{10}H_{10}$** must also be the molecular formula.

CHAPTER THREE

3.19   a. $C_3H_8 + 5O_2 \rightarrow 3CO_2 + 4H_2O$
       b. $Fe_2O_3 + 3CO \rightarrow 2Fe + 3CO_2$
       c. $Mg_3N_2 + 6H_2O \rightarrow 2NH_3 + 3Mg(OH)_2$
       d. $Ca_3(PO_4)_2 + 2H_2SO_4 \rightarrow 2CaHPO_4 + Ca(HSO_4)_2$
       e. $3K_2CO_3 + Al_2Cl_6 \rightarrow Al_2(CO_3)_3 + 6KCl$
       f. $8KClO_3 + C_{12}H_{22}O_{11} \rightarrow 8KCl + 12CO_2 + 11H_2O$
       g. $KOH + H_3PO_4 \rightarrow KH_2PO_4 + H_2O$

3.21   a. $3Fe + 4H_2O \rightarrow Fe_3O_4 + 4H_2$
       b. (1 g Fe)(55.85 g/1 mol) = 1.80×10^{-2} mol Fe
          $(1.80 \times 10^{-2}$ mol Fe)(4 mol H_2/3 mol Fe) = **2.40×10^{-2} mol H_2**
       c. 0.170(60.0 g) = 10.2 g Fe
          (10.2 g Fe)(1 mol/55.85 g) = 0.183 mol Fe
          (0.183 mol Fe)(1 mol Fe_3O_4/3 mol Fe) = 0.0610 mol Fe_3O_4
          (0.0610 mol Fe_3O_4)(231.5 g/1 mol) = **14.1 g Fe_3O_4**

3.23   a. $2C_8H_{18} + 25O_2 \rightarrow 16CO_2 + 18H_2O$
       b. (1 g C_8H_{18})(1 mol/114.2 g) = 8.76×10^{-3} mol C_8H_{18}
          $(8.76 \times 10^{-3}$ mol C_8H_{18})(16 mol CO_2/2 mol C_8H_{18}) = $(7.00 \times 10^{-2}$ mol CO_2)
          $(7.00 \times 10^{-2}$ mol CO_2)(44.01 g/1 mol) = **3.08 g CO_2**
       c. (124 g C_8H_{18})(1 mol/114.2 g) = 1.09 mol C_8H_{18}
          (32 g O_2)(1 mol/32.00 g) = 1.0 mol O_2
          O_2 must be the limiting reactant since there is much less than the required 25:2 mole ratio required to react with all of the *iso*-octane.
          After Reaction: all O_2 is used up, 0.64 mol CO_2 is produced, 0.72 mol H_2O produced, and 1.01 mol C_8H_{18} remain.
          You are left with **115 g C_8H_{18}, 28 g CO_2 and 13 g H_2O**.

3.25   Write the balanced reaction: $CS_2 + 3O_2 \rightarrow CO_2 + 2SO_2$
       (1.00 L CS_2)(10^3 mL/L)(1.26 g/L)/76.13 g/mol = 16.55 mol CS_2
       (16.55 mol CS_2)(2 mol SO_2/mol CS_2)(64.06 g/mol) = **2120 g SO_2**

3.27   a. $4NH_3 + 5O_2 \rightarrow 4NO + 6H_2O$
          $2NO + O_2 \rightarrow 2NO_2$
          $3NO_2 + H_2O \rightarrow 2HNO_3 + NO$
          $HNO_3 + NH_3 \rightarrow NH_4NO_3$

CHAPTER THREE

    b. **Two moles N** are required for every one mole NH_4NO_3.

    c. $(1000.\text{ g }HNO_3)(1\text{ mol}/63.01\text{ g}) = 15.87$ moles HNO_3

    In this sequence of steps, the only nitrogen-containing reactant is ammonia; the other nitrogen containing compounds involved (NO and NO_2) are intermediates, both formed and consumed in the series of reactions. Therefore, all of the nitrogen atoms in HNO_3 ultimately came from NH_3. There must be a 1:1 mole ratio between HNO_3 and NH_3.

    In order to produce 15.87 mol HNO_3, 15.87 mol NH_3 are required.

    $(15.87\text{ mol }NH_3)(17.03\text{ g}/1\text{ mol}) = $ **270.3 g NH_3**

3.29 $M + 2X \rightarrow MX_2$ where M = metal, X = nonmetal

    a. For 5.00 g of X, $(5.00\text{ g X})(1\text{ mol}/80.\text{ g}) = 0.0625$ mol X

    $(0.0625\text{ mol X})(1\text{ mol M}/2\text{ mol X}) = 0.03125$ mol M

    $(0.03125\text{ mol M})(24\text{ g/mol}) = 0.75$ g M

    0.75 g of metal are required.

    b. $(0.03125\text{ mol M})(1\text{ mol }MX_2/1\text{ mol M}) = 0.03125$ mol MX_2

    $(0.03125\text{ mol }MX_2)(184\text{ g}/1\text{ mol}) = 5.75$ g MX_2

    So **5.75 g MX_2** will be formed.

3.31 $(25.0\text{ g KClO})(1\text{ mol}/90.55\text{ g}) = 0.276$ mol KClO

    $(0.276\text{ mol KClO})(1\text{ mol }Cl_2/1\text{ mol KClO}) = 0.276$ mol Cl_2 required

    $(0.276\text{ mol }Cl_2)(1\text{ mol }MnO_2/1\text{ mol }Cl_2) = 0.276$ mol MnO_2 required

    $(0.276\text{ mol }MnO_2)(86.94\text{ g}/1\text{ mol}) = $ **24.0 g MnO_2 required**.

3.33   a. 1.00 mol HCl can produce **0.500 mol H_2** and **0.500 mol Cl_2**

    1.00 mol H_2O can produce **1.00 mol H_2** and **0.500 mol O_2**

    1.00 mol NH_3 can produce **1.50 mol H_2** and **0.500 mol N_2**

    b. 1.00 mol Cl_2 requires 1.00 mol $H_2 = 2.02$ g H_2

    1.00 mol O_2 requires 2.00 mol $H_2 = 4.03$ g H_2

    1.00 mol N_2 requires 3.00 mol $H_2 = 6.05$ g H_2

3.35   a. $3LiAlH_4 + 4BF_3 \rightarrow 3LiF + 3AlF_3 + 2B_2H_6$

    b. $(100.\text{ g }LiAlH_4)(1\text{ mol}/37.95\text{ g}) = 2.64$ mol $LiAlH_4$

    $(225\text{ g }BF_3)(1\text{ mol}/67.81\text{ g}) = 3.32$ mol BF_3

    BF_3 is the limiting reagent because there is not enough of it to react with 2.64 mol $LiAlH_4$ (3.52 mol BF_3 would be required).

    Theoretically, 3.32 mol BF_3 should produce 1.66 mol B_2H_6.

    Theoretical yield = $(1.66\text{ mol }B_2H_6)(27.67\text{ g}/1\text{ mol}) = $ **45.9 g B_2H_6**

CHAPTER THREE

3.37 $4HCl + O_2 \rightarrow 2H_2O + 2Cl_2$

    a. **20.0 moles HCl** are needed to produce 10.0 moles Cl_2.

    b. **5.00 moles Cl_2** are produced from 10.0 moles HCl and excess O_2.

    c. When 10.0 moles HCl are mixed with 10.0 moles O_2, HCl is the limiting reactant, and **5.00 moles Cl_2** are produced.

3.39 All of the carbon present in the original compound will end up in the carbon dioxide:
$(3.210\ g\ CO_2)(1\ mol\ CO_2/44.01\ g\ CO_2) = 0.07294\ mol\ CO_2 = 0.07294\ mol\ C$

Similarly, all of the hydrogen present in the original compound will end up in the water:
$(1.751\ g\ H_2O)(1\ mol\ H_2O/18.02\ g\ H_2O) = 0.09717\ mol\ H_2O$

Since there are 2 hydrogens in each H_2O, $2(0.09717) = 0.1943\ mol\ H$

The mole ratio of H:C = $(0.1943\ mol/0.07294\ mol) = 2.664$, which is very nearly 8:3.
The empirical formula of the compound is **C_3H_8**.

3.41 All of the carbon present in the original compound will end up in the carbon dioxide:
$(14.7\ g\ CO_2)(12.01\ g\ C/44.01\ g\ CO_2) =$ **4.01 g C**

Similarly, all of the hydrogen present in the original compound will end up in water:
$(6.0\ g\ H_2O)(2.016\ g\ H/18.02\ g\ H_2O) =$ **0.67 g H**

Finally, all of the original 10.00 g mass not yet accounted for must have been oxygen:
$10.00\ g - 4.01\ g - 0.67\ g =$ **5.32 g O**

Now use these masses to determine the mole ratio of H:C:O in the original compound.
$(4.01\ g\ C)(1\ mol/12.01\ g) = 0.333\ mol\ C$
$(0.67\ g\ H)(1\ mol/1.008\ g) = 0.66\ mol\ H$
$(5.32\ g\ O)(1\ mol/16.00\ g) = 0.333$

This = 1:2:1 within experimental error. The empirical formula is **CH_2O**.

3.43 $(2.000\ g\ AgBr)(1\ mol/187.77\ g) = 0.01065\ mol\ AgBr = 0.01065\ mol\ Ag$ in original sample
$(0.01065\ mol\ Ag)(107.87\ g/1\ mol) = 1.149\ g\ Ag$ in original sample
$(1.149\ g\ Ag)/(2.000\ g\ sample) = 0.5745$
57.45% Ag in the alloy

3.45 First determine the number of moles of Ag_2SO_4:
$(1.992\ g\ Ag_2SO_4)(1\ mol/311.80\ g) = 6.389 \times 10^{-3}$ moles Ag_2SO_4.
$(6.389 \times 10^{-3}\ moles\ Ag_2SO_4)(1\ mol\ SO_4^{2-}/1\ mol\ Ag_2SO_4)$
= total moles $SO_4^{2-} = 6.389 \times 10^{-3}$ moles SO_4^{2-} = moles of K_2SO_4 + moles of Na_2SO_4
1.000 g = mass of K_2SO_4 + mass of Na_2SO_4
Let x = mass of K_2SO_4, $(1 - x)$ = mass of Na_2SO_4
$(x/174.25\ g/mol) + (1 - x)/(142.04\ g/mol) = 6.389 \times 10^{-3}$ moles

CHAPTER THREE

Multiply both sides by (174.25)(142.04):
142.04 x + 174.25 − 174.25 x = 158.1
x = 0.500 g, (1 − x) = 0.500 g

The original mixture was **50.0% K_2SO_4 and 50.0% Na_2SO_4**.

3.47 2.15 g AgCl (1 mol/143.32 g) = 0.0150 mol AgCl = 0.0150 mol Cl^- which came from the original NaCl and/or KCl

Let x = original mass of KCl, (1.00 − x) = original mass of NaCl

Then (x/74.55 g/mol) + (1 − x)/(58.44 g/mol) = 0.0150 moles. Now solve for x:

x = 0.571 g KCl in original mixture, 1.00 − x = 0.43 g NaCl in original mixture

The original sample was a **mixture of 0.43 g NaCl and 0.57 g KCl.**

3.49 We know that the sum of the masses must be 1.00 g:

$m_{Cu_2O} + m_{CuO}$ = 1.00 g.

The total moles of copper must be (0.839 g Cu)(1 mol/63.55 g) = 0.0132 mol.

2(moles Cu_2O) + (moles CuO) = 0.0132 mol

Let x = m(Cu_2O); then (1.00 − x) = m(CuO)

2(x/143.1 g/mol) + (1.00 − x)/(79.55 g/mol) = 0.0132 mol

Multiply both sides by the product of molar masses and solve for x:

159.1x + 143.1 − 143.1x = 150.3;

x = 0.45 g Cu_2O; (1.00 − x) = 0.55 g CuO

The original mixture was **45% Cu_2O and 55% CuO**.

3.51 From the information given, we know that for every 1.00 g of Fe in the original mixture, there is 1.00 g of S. Then

(1.00 g Fe)(1 mol/55.85 g) = 0.0179 mol Fe

(1.00 g S)(1 mol/32.06 g) = 0.0312 mol S

The iron will be used up first, so iron is the limiting reactant, and **sulfur is in excess**. Since Fe and S combine in a 1:1 mole ratio in this reaction, 0.0179 mol S will be used up, leaving (0.0312 mol − 0.0179 mol) = 0.0133 mol unreacted sulfur.

(0.0133 mol S)(32.06 g/1 mol) = 0.426 g sulfur remain.

Of the original 2.00 g sample, 0.426 g are left unreacted.

(0.426 g)/(2.00 g) = **0.213 = 21.3% unreacted material**

This answer does not depend on the original mass used for both iron and sulfur. Try the calculation using 0.500 g Fe and 0.500 g S to show that this is true.

CHAPTER THREE

3.53 Every 100.0 g of the acid contains 1.04 g H, 33.04 g S, and 65.92 g O.
(1.04 g H)(1 mol/1.008 g) = 1.03 mol H
(33.04 g S)(1 mol/32.06 g) = 1.030 mol S
(65.92 g O)(1 mol/16.00 g) = 4.120 mol O
The ratio H:S:O is 1:1:4, so the **empirical formula is HSO_4**.
Molecular formula = n(HSO_4), where n = molar mass/(empirical formula mass).
n = (194/97.06) = 2; The **molecular formula is $(HSO_4)_2$ or $H_2S_2O_8$**.

3.55 a. The balanced equation is $C(s) + CO_2(g) \rightarrow 2CO(g)$

b. Based on the reaction
$CH_4 + H_2O \rightarrow CO + 3H_2$:
(10.0 g CH_4)(1 mol/16.04 g) = 0.623 mol CH_4
(10.0 g H_2O)(1 mol/18.02 g) = 0.555 mol H_2O
H_2O is the limiting reactant, and CH_4 is in excess.

c. The theoretical yield of the reaction in part b is 0.555 mol CO (based on the 0.555 mol of H_2O, the limiting reactant).
(0.555 mol CO)(28.01 g/1 mol) = 15.5 g CO = theoretical yield
Actual yield = (10.0 g CO)/(15.5 g CO) = 0.645 = **64.5%**

d. (10.0 g CO)(1 mol/28.01 g) = 0.357 mol CO
(0.357 mol CO)(3 mol H_2/1 mol CO) = 1.07 mol H_2
(1.07 mol H_2)(2.016 g/1 mol) = **2.16 g H_2**

3.57. a. Let the first reaction proceed to the extent x: 2x mol ClO_2 and x mol Cl_2 are produced.
Let the second reaction proceed to the extent y: 3y mol Cl_2 produced.
(moles ClO_2) = (moles Cl_2); 2x = x + 3y, y/x = 1/3
1/(1 + 3) = 0.25; **25% $NaClO_3$** is undergoing the secondary reaction.

b. $(1.0 \times 10^3$ g $NaClO_3)$(1 mol/106.44 g) = 9.4 mol $NaClO_3$ reacts
(0.75)(9.4 mol $NaClO_3$) = 7.0 mol $NaClO_3$ will undergo primary reaction
(0.25)(9.4 mol $NaClO_3$) = 2.4 mol $NaClO_3$ will undergo secondary reaction
This will produce
7.0 mol ClO_2 (4.8×10^2 g) and 7.0 mol Cl_2 (5.0×10^2 g).

3.59 The numbers of moles are:
34.5 g Na/23.0 g/mol = 1.50 mol Na
13.5 g Al/27.0 g/mol = 0.500 mol Al
(66.8 g − 13.5 g)/35.45 g/mol = 1.50 mol Cl

Al and Cl are present in a 1:3 ratio
Empirical Formula = **$AlCl_3$**; simplest equation: **$AlCl_3 + 3Na \rightarrow Al + 3NaCl$**

CHAPTER THREE

3.61 a. $KClO_4 + 4C \rightarrow KCl + 4CO$
$KClO_4 + 2C \rightarrow KCl + 2CO_2$

b. Assume only the second reaction is occuring to any significant extent under these conditions.
$(1.0 \times 10^3 \text{ g } KClO_4)(1 \text{ mol}/138.55 \text{ g}) = 7.2 \text{ mol } KClO_4$
$(7.2 \text{ mol } KClO_4)(2 \text{ mol C}/1 \text{ mol } KClO_4) = 14 \text{ mol C required.}$
$(14 \text{ mol C})(12.011 \text{ g}/1 \text{ mol}) = \textbf{1.7} \times \textbf{10}^\textbf{2} \textbf{ g C}$

3.63 1 lb = 454 g and 1 mol of $KHCO_3$ = 100.1 g.
$(1000. \text{ lb})(454 \text{ g/lb})(1 \text{ mol}/100.1 \text{ g}) = 4.54 \times 10^3 \text{ mol}$
A gaylord of $KHCO_3$ contains $\textbf{4.54} \times \textbf{10}^\textbf{3}$ **mol $KHCO_3$**.

3.65 The molar masses are: M_{CS_2} = 76.13 g/mol; M_{Cl_2} = 70.90 g/mol;
M_{CCl_4} = 153.81 g/mol; $M_{S_2Cl_2}$ = 135.02 g/mol

Since the **products have a greater total mass per mole (288.83 g) than the reactants (147.03 g),** the equation is not balanced as written. Otherwise, the reaction would produce mass which is in violation of the law of conservation of mass. Of course, you could also figure out that the reaction is not balanced by recognizing that there are different numbers of Cl atoms on the two sides of the equation.

We can balance the reaction by inspection:
$CS_2 + 3Cl_2 \rightarrow CCl_4 + S_2Cl_2$.
Now the total mass on the reactant side is:
$(1 \text{ mol})(76.13 \text{ g/mol}) + (3 \text{ mol})(70.90 \text{ g/mol}) = \textbf{288.83 g}$
And the total mass on the product side is:
$(1 \text{ mol})(153.81 \text{ g/mol}) + (1 \text{ mol})(135.02 \text{ g/mol}) = \textbf{288.83 g}$
Thus, material balance (mass conservation) follows directly from a balanced equation.

3.67 a. $(1 \times 10^3 \text{ g } Fe_2O_3)(1 \text{ mol}/159.7 \text{ g}) = 6.262 \text{ mol } Fe_2O_3$
$(6.262 \text{ mol } Fe_2O_3)(3 \text{ mol C}/1 \text{ mol } Fe_2O_3) = 18.79 \text{ mol C}$
$(18.79 \text{ mol C})(12.011 \text{ g}/1 \text{ mol}) = \textbf{225.7 g C required}$

b. $(18.79 \text{ mol C})(3 \text{ mol CO}/3 \text{ mol C})(1 \text{ mol } O_2/2 \text{ mol CO}) = 9.395 \text{ mol } O_2$
$(9.395 \text{ mol } O_2)(32.00 \text{ g}/1 \text{ mol}) = \textbf{300.6 g } \textbf{O}_\textbf{2}$ **required.**

CHAPTER THREE

3.69 The molar masses of the three compounds are:
$CaWO_4$: 287.93 g/mol; $FeWO_4$: 303.70 g/mol; Na_2WO_4: 293.83 g/mol

Each compound will be represented by its initial letter. We are given that:
$m_N = 1.0$ g and $m_C + m_F = 1.0$ g

All the tungstate appears in the Na_2WO_4, so we can write for the total number of moles of tungstate
$n_N = n_C + n_F$
1.0 g/293.83 g/mol = m_C/287.93 + m_F/303.70 = m_C/287.93 + (1.0 − m_C)/303.70.

Clear fractions by multiplying through by the product of all three molar masses:
1.0(287.93)(303.70) = m_C(293.83)(303.70) + 1.0(293.83)(287.93) − m_C(293.83)(287.93)

87444 = 89236 m_C + 84602 − 84602 m_C

m_C = (87444 − 84602)/(89236 − 84602) = 2842/4634 = 0.61 g

% $CaWO_4$ = (0.61/1.0) × 100 = **61%**

3.73 a. The relevant molar masses are
Cumene: 120.02 g/mol; phenol: 94.11 g/mol; acetone: 58.08 g/mol.
If 600. × 10^6 lb phenol are produced, that is
(600. × 10^6 lb)(454 g/lb)/(94.11 g/mol) = 2.89 × 10^9 mol.
If 350 × 10^6 lb acetone are produced, that is (350 × 10^6 lb)(454 g/lb)/(58.08 g/mol)
= 2.73 × 10^9 mol

We will use this as the basis for our "minimum" consumption of cumene.
1 mole of cumene is needed for each mole of phenol produced; therefore, at least 2.89 × 10^9 moles of cumene were used.

Pound per hour minimum world scale consumption
= (2.89 × 10^9 mol)(120.019 g/mol)(1 pound/454 g)(1/0.10 amount of world consumption)/[(365 days/year)(24 hours/day)]
= **8.7 × 10^5 lb/hr** world consumption.

b. Phenol is produced in greater yield.

CHAPTER FOUR

Gases

4.1 The relationship between the atmospheric pressure P and the height h of a column of liquid that can be supported by the pressure is P = dgh. In this problem, it is probably easiest to convert all quantities to SI units, which will yield the column height in meters.
P = (5.0 atm)(1.01 × 10^5 Pa/atm) = 5.05 × 10^5 Pa = 5.05 × 10^5 kg/(m·s^2)
d = 1.00 × 10^3 kg/m^3 and g = 9.81 m/s^2.
P = 5.05 × 10^5 kg/m·s^2 = (1.00 × 10^3 kg/m^3)(9.81 m/s^2) h.
h = **52 m**.

4.3 The height h of a column of liquid supported by a pressure P is given by P = dgh. Conversion of all quantities to SI units will yield h in meters.
P = (785/760 atm) (1.01 × 10^5 Pa/atm) = 1.04 × 10^5 Pa = 1.04 × 10^5 kg/(m·s^2),
d = 789 kg/m^3, and
g = 9.81 m/s^2.
P = 1.04 × 10^5 kg/m·s^2 = (789 kg/m^3) (9.81 m/s^2) h.
h = **13.4 m**.

4.5 In an open-ended manometer, the height difference between the two legs of the U-tube is equal to the pressure difference between the gas and the atmosphere: ΔP = P$_{atm}$ − P$_{gas}$. You know that P$_{gas}$ < P$_{atm}$ because the level of the Hg is higher adjacent to the gas container.
ΔP = 33.1 mm Hg = 748.4 mm − P$_{gas}$
P$_{gas}$ = 748.4 − 33.1 = 715.3 mm Hg = **715.3 torr**.

4.7 In an open-ended manometer, the height difference between the two legs of the U-tube equals the pressure difference between the gas and the atmosphere. In this case, P$_{gas}$ > P$_{atm}$ because the Hg is higher in the leg adjacent to the atmosphere.
ΔP = P$_{gas}$ − P$_{atm}$ = 22 cm − 8 cm = 14 cm = 140 mm Hg = 140 torr
P$_{gas}$ = P$_{atm}$ + 140 torr (this number contains only two significant figures)
If P$_{atm}$ = 1.00 atm = 760 torr, then P$_{gas}$ = 900 torr = 900/760 = **1.2 atm**
If P$_{atm}$ = 1.00 kPa = 7.53 torr, then P$_{gas}$ = 147.5 torr = 147.5/760 = 0.194 atm = **20. kPa**

4.9 Use Boyle's law in the form P$_i$V$_i$ = P$_f$V$_f$ (temperature is constant) and take care that units of P and V are consistent on both sides of the equation.
1. (1.00 atm)(22.4 L) = P$_f$(2.24 L)   P$_f$ = **10.0 atm**
2. P$_i$(10.0 mL) = (101.3 kPa)(1.00 × 10^3 mL)   P$_i$ = **1.01 × 10^4 kPa**
3. (100. Pa)(V$_i$) = (1.00 × 10^5 Pa)(1.00 L)   V$_i$ = **1.00 × 10^3 L = 1.00 m^3**
4. (2.5 atm)(2.50 L) = P$_f$(0.0250 L)   P$_f$ = **2.5 × 10^2 atm**

CHAPTER FOUR

4.11 Using the ideal gas equation $PV = nRT$ and $n = m/M$, we can relate the density of a gas to its pressure by $d = PM/RT$. Using grams for molar mass will yield the density in grams/L.
$d = PM/RT = (1.00\text{ atm})(32.00\text{ g/mol})/[(0.08206\text{ L·atm/mol·deg})(273\text{K})] = \mathbf{1.43\text{ g/L}}.$

4.13 From the ideal gas equation, we write $PV = nRT = (m/M)RT$. The ratio m/M must contain the same units; we will use grams. Then
$(1.55\text{ atm})(3.12\text{ L}) = (m/32.00\text{ g/mol})(0.08206\text{ L·atm/mol·K})(298\text{K})$
$m = \mathbf{6.33\text{ g }O_2}$

4.15 We use $P_iV_i/T_i = P_fV_f/T_f$ for a fixed amount of gas. Remember that temperature must be in kelvin units. In this case, $P_i = P_f$, and so the ratio becomes
$1818\text{ mL}/298\text{K} = V_f/273\text{K}$
$V_f = \mathbf{1665\text{ mL}}$

4.17 Use $P_iV_i/T_i = P_fV_f/T_f$. Units must be the same on both sides of the equation.
$(0.785\text{ atm})(10.0\text{ L})/350\text{K} = (0.656\text{ atm})(15.0\text{ L})/T_f$
$T_f = \mathbf{439\text{K}}$

4.19 First, calculate the number of moles of gas particles from the ideal gas equation $PV = nRT$.
$P = (1.00 \times 10^{-6}\text{ torr})(1\text{ atm}/760\text{ torr}) = 1.32 \times 10^{-9}\text{ atm}$
$n = PV/RT = (1.32 \times 10^{-9}\text{ atm})(1.00\text{ L})/[(0.08206\text{ L·atm/mol·K})(273\text{K})]$
$= 5.89 \times 10^{-11}\text{ moles}$
Then use $N = nN_A = (5.89 \times 10^{-11}\text{ moles})(6.022 \times 10^{23}\text{ particles/mol})$
$N = \mathbf{3.55 \times 10^{13}\text{ particles}}$

4.21 $(200.\text{ g }CaCO_3)/100.09\text{ g/mol} = 2.00\text{ mol }CaCO_3$, which will also produce 2.00 mol CO_2. Calculate the volume from the ideal gas equation with $P = 750/760 = 0.987\text{ atm}$.
$V = nRT/P = (2.00\text{ mol})(0.08206\text{ L·atm/mol·K})(1273\text{K})/(0.987\text{ atm})$
$V = \mathbf{212\text{ L}}$

4.23 Using the ideal gas equation $PV = nRT$ and $n = m/M$, we can relate the pressure of a gas to its density by
$P = (m/V)(RT/M) = dRT/M$
$d = PM/RT = (1.00\text{ atm})(153.8\text{ g/mol})/[(0.08206\text{ L·atm/mol·K})(373\text{K})]$
$d = \mathbf{5.02\text{ g/L}}$

4.25 Density and molar mass of an ideal gas are related by $d = PM/RT$.
$M = dRT/P = (8.06\text{ g/L})(0.08206\text{ L·atm/mol·K})(298\text{K})/(1.50\text{ atm})$
$M = \mathbf{131\text{ g/mol}}$

CHAPTER FOUR

4.27 The relationship between pressure, density and temperature for an ideal gas is
$P = dRT/M$
This problem can be solved by a simple ratio:
$P_f/P_i = (d_f/d_i)(T_f/T_i)$, where $d_f = d_i$ and T must be in kelvin units.
$P_f/(1.000 \text{ atm}) = (298/273)$
$P_f = \mathbf{1.09 \text{ atm}}$

4.29 First, determine the empirical formula from the composition data given. In every 100.00 g of the compound, there are 24.26 g C, 4.08 g H and 71.66 g Cl. The numbers of moles are:
(24.26 g C)(1 mol/12.01 g) = 2.020 mol C
(4.08 g H)(1 mol/1.008 g) = 4.05 mol H
(71.66 g Cl)(1 mol/35.45 g) = 2.021 mol Cl
Dividing each number by 2.02 to get the smallest whole-number ratios of the elements gives the empirical formula CH_2Cl, which has a molar mass of 49.5 g/mol. Next, we determine the true molar mass from the ideal gas equation: $M = dRT/P$.
d = (0.4416 g)/(0.1402 L) = 3.150 g/L
P = (740 torr)(1 atm/760 torr) = 0.974 atm
M = (3.150 g/L)(0.08206 L·atm/mol·K)(373K)/(0.974 atm) = 99.0 g/mol
This is twice the molar mass of the empirical formula CH_2Cl, so the true formula of the compound is $(CH_2Cl)_2$, or $\mathbf{C_2H_4Cl_2}$.

4.31 Pressure is related to density and molar mass by $P = dRT/M$, from which the density can be calculated:
d = PM/RT = (1.000 atm)(44.01 g/mol)/[(0.08206 L·atm/mol·K)(273K)]
d = **1.96 g/L**

4.33 Density is related to pressure and molar mass by $d = PM/RT$ which shows that density is proportional to M/T if the pressure is constant. Take the ratio of densities for the two gases:
$(d_{air})/(d_{He}) = (M_{air}/T_{air}) \times (T_{He}/M_{He})$
For air and He to have the same lifting power, they must have the same density, so we can write
1 = (28.8 g/mol)/(T_{air}) × (298K)/(4.003 g/mol)
T_{air} = (28.8/4.003)298 = **2140K**

CHAPTER FOUR

4.35 For a mixture of ideal gases, $P_T = p_A + p_B + p_C + ...$ and each partial pressure p_i is $X_i P_T$ (Dalton's Law). Each mol fraction X_i is the percent by volume (or by moles) divided by 100. Thus,
$X_{N_2} = 0.78$, $X_{O_2} = 0.21$ and $X_{Ar} = 0.01$
$p_{N_2} = (0.78)(772 \text{ torr}) =$ **602 torr**
$p_{Ar} = (0.01)(772 \text{ torr}) =$ **8 torr**
The remainder of the total pressure is oxygen.

4.37 In 100. mol of air, there are 78 mol N_2, 21 mol O_2 and 1 mol Ar. The masses of each gas in 100. mol are:
  (78 mol N_2)(28.01 g/mol) = 2.2×10^3 g N_2 = 2.2 kg N_2
  (21 mol O_2)(32.00 g/mol) = 6.7×10^2 g O_2 = 0.67 kg O_2
  (1 mol Ar)(39.95 g/mol) = 40 g Ar = 0.04 kg Ar
  Total mass = 2.9 kg
% N_2 = $(m_{N_2}/m_T) \times 100$ = (2.2 kg/2.9 kg) × 100 = **76% by mass**
% O_2 = (0.67 kg/2.9 kg) × 100 = **23% by mass**
% Ar = (0.04 kg/2.9 kg) × 100 = **1% by mass**
Average molar mass = 2900 g/100 moles = **29 g/mol**

4.39 For an ideal gas, the % by volume is the same as % by moles. Therefore, each mol fraction is the mole percent divided by 100. By Dalton's Law, each partial pressure is $X_i P_T$ and the total pressure, 1.00 atm in this case, is the sum of the individual partial pressures.
  a. $p_{O_2} = (0.21)(1.00 \text{ atm}) =$ **0.21 atm**
  b. $p_{N_2} = (0.78)(1.00 \text{ atm}) =$ **0.78 atm**
  c. if all the oxygen is removed, the total pressure is simply that of N_2 and Ar:
    $P_T = 1.00 - 0.21 =$ **0.79 atm**

4.41 In every 100 mol of gas, there are 20.0 mol of CH_4, 5.0 mol of C_2H_6 and 75.0 mol of CO_2. The individual masses are
  (20.0 mol CH_4)(16.04 g/1 mol CH_4) = 321 g CH_4
  (5.0 mol C_2H_6)(30.07 g/1 mol C_2H_6) = 150 g C_2H_6
  (75.0 mol CO_2)(44.01 g/1 mol CO_2) = 3300 g CO_2
  Total mass = 3770 g
% CH_4 = (321 g/3770 g) × 100 = **8.51%**
% C_2H_6 = (150 g/3770 g) × 100 = **4.0%**
% CO_2 = (3300 g/3770 g) × 100 = **87.5%**

CHAPTER FOUR

4.43 First, write the balanced equation for the combustion using the variables m and n:
$C_nH_m + (n + 0.25m)O_2 \rightarrow nCO_2 + 0.50mH_2O$
We assume there is no O_2 remaining the product gas mixture. Next, calculate the molfraction of H_2O in the mixture and set it equal to the mol ratio n_{H_2O}/n_T:
$X_{H_2O} = n_{H_2O}/n_T = 0.50 \text{ m}/(n + 0.50m) = 0.686 \text{ atm}/1.200 \text{ atm} = 0.572$.
Now solve this equation for the ratio of H to C, m/n, in the hydrocarbon:
$0.50m = 0.572n + 0.286m$; $0.214m = 0.572n$; $m/n = 2.67 = 8/3$
The empirical formula is **C_3H_8**.

4.45 The translational kinetic energy of n moles of an ideal gas is $KE_{tr} = 3/2 \, nRT$.
$KE_{tr} = 3/2(1 \text{ mole})(8.314 \text{ J/mol·K})(298K) = 3.72 \times 10^3$ J
$KE_{tr} =$ **3.72 kJ**

4.47 The root-mean-square speed of a gas is $v_{rms} = \sqrt{3RT/M}$.
In order to have units of m/s for v, we will use SI units throughout:

$$v_{rms} = \sqrt{3(8.314 \text{ J/mol·deg})(300K)/(4.00 \times 10^{-3} \text{ kg})}$$

$v_{rms} =$ **1.37×10^3 m/s**

4.49 a. Since the two gases are at the same temperature, their average molar kinetic energies must be the same.
b. According to Avogadro's law, at constant T and P, the volume of a gas is directly proportional to the number of particles present. Since $V_B = 2V_A$, there must be twice as many B particles as A particles.
c. As noted in part (a), the two gases have the same molar translational kinetic energy. However, they could be moving at different average speeds because of their different molar masses. At any temperature, the average speed varies inversely with the square root of molar mass:

$$v_A/v_B = \sqrt{M_B/M_A}$$

Since both molar masses are about 44.0 g/mol, the two gases will have nearly identical average speeds.

4.51 For equal numbers of gas particles A and B at the same temperature, each rate of effusion will be inversely proportional to the square root of molar mass. Neon will effuse faster because it is lighter.

$\text{Rate}_{Ne}/\text{Rate}_{Ar} = \sqrt{M_{Ar}/M_{Ne}} = \sqrt{39.95/20.018}$
= **1.41**

CHAPTER FOUR

4.53 For equal numbers of gas particles A and B at the same temperature, each rate of effusion will be inversely proportional to the square root of its molar mass.

$$\text{Rate}_{PH_3}/\text{Rate}_{NH_3} = \sqrt{M_{NH_3}/M_{PH_3}} = \sqrt{17.03/34.00} = 0.708$$

∴ Rate$_{PH3}$ = 0.708(8.02 cm^3/s) = **5.68 cm^3/s**

4.55 Rearrange the van der Waals equation and solve for pressure:
   P = nRT/(V − nb) − n^{2}a/V^2

For each gas, n = 10.0 mol, V = 15.0 L, and T = 250K, and the van der Waals constants are given in Table 4.4.
   For He, a = 0.03412 L^2·atm/mol^2, b = 0.02370 L/mol
   P = (10.0)(0.08206)(250K)/(15.0 − 0.2370) − 3.412/225

P = 13.9 atm
   For Cl$_2$, a = 6.493 L^2·atm/mol^2, b = 0.05622 L/mol
   P = (10.0)(0.08206)(250K)/(15.0 − 0.5622) − 649.3/225

P = 11.3 atm
Less pressure is required to confine 10 moles of Cl$_2$ within a given volume than is required for He, even though Cl$_2$ is a larger molecule than He. This is because of the relatively strong attractive forces between Cl$_2$ molecules, reflected in its large "a" value.

4.57 We want to write an expression for the compressibility factor Z = PV/nRT based on the van der Waals equation (P + n^{2}a/V^2)(V − nb) = nRT.

Begin by multiplying out the left-hand side of the equation
   PV = n^{2}aV − nbP − n^{3}ab/V^2 = nRT

Next, divide through by nRT:
   PV/nRT + na/VRT − bP/RT − n^{2}ab/V^{2}RT = 1

and rearrange, setting PV/nRT = Z
   Z = 1 + (bP − na/V + n^{2}ab/V^2)/RT.

The term following the "1" collectively represents the deviations from ideal behavior ("Z = 1"). We see that Z will be tend to be less than unity (corresponding to negative derivations) for
- small molecular size (small van der Waals "b")
- strong intermolecular attractions (large "a")
- low temperature (maximizes the deviations from ideality)
- moderately low pressure

4.59 Solve by Charles' Law:
   V$_1$/T$_1$ = V$_2$/T$_2$, so V$_2$ = V$_1$(T$_2$/T$_1$)
   V$_2$ = (0.110 m^3)(400.K/273K) = **0.161 m^3**

CHAPTER FOUR

4.61 Solve by Charles' Law
$V_1/T_1 = V_2/T_2$, so $V_2 = V_1(T_2/T_1)$
$V_2 = (2.0 \text{ L})(77\text{K}/273\text{K}) = $ **0.56 L**

4.63 Using the ideal gas equation $PV = nRT$ and $n = m/M$, we can determine the molar mass of ozone from $M = mRT/PV$.
$m = 6.7624 - 6.5998 = 0.1626$ g
$P = (274.4 \text{ torr})(1 \text{ atm}/760 \text{ torr}) = 0.3610$ atm
$V = 0.23567$ L
$M = (0.1626 \text{ g})(0.08206 \text{ L·atm/mol·K})(301.4\text{K})/[0.3610 \text{ atm})(0.23567 \text{ L})]$
$M = $ **47.27 g/mol**

4.65 The relation between density, pressure and molar mass for an ideal gas is $P = dRT/M$. Since we are comparing two gases we can use the ratios of quantities:
$P_X/P_Y = (d_X M_Y)/(d_Y M_X)$
We are given that $M_Y = 2M_X$ and $d_Y = d_X/2$. Therefore, $P_X/P_Y = (2)(2) = 4$.
The pressure of X is **4 times** that of Y.

4.67 a. We will first use the results of the data for oxygen to determine the volume of the bulb. Write the ideal gas equation $V = nRT/P = mRT/MP$
$V = (2.4 \text{ g})(0.08206 \text{ L·atm/mol·K})(298.2\text{K})/[(32.00 \text{ g/mol})(1.00 \text{ atm})]$
$= 1.84$ L (this has just two significant figures)
Now solve for M of the unknown gas: $M = mRT/PV$
$M = (7.3 \text{ g})(0.08206 \text{ L·atm/mol·K})(298.2\text{K})/[(1.00 \text{ atm})(1.84 \text{ L})]$
$M = $ **97 g/mol**
b. The density is $m/V = 7.3 \text{ g}/1.84 \text{ L} = $ **4.0 g/L**.

4.69 a. The number of moles of SO_2 is **four times** that of He.
Reason: $n = PV/RT$; at constant temperature, $(PV)_{SO_2}$ is $4(PV)_{He}$.
b. The average kinetic energy of SO_2 molecules is the **same** as that of He atoms.
Reason: $KE = 3/2 \, kT$, and the temperatures are the same.
c. The average speed of SO_2 molecules is **one-fourth** that of He atoms.
Reason: average speed v is inversely proportional to $\sqrt{M}$.
$v_{SO_2}/v_{He} = \sqrt{4.0/64} = 1/4$
d. He diffuses at **twice the rate** of SO_2. Reason: diffusion rate is proportional to $P \cdot v_{rms}$, and v_{rms} is inversely proportional to $1/\sqrt{M}$.
$\text{rate}_{He}/\text{rate}_{SO_2} = (1/2)(4/1) = 2$.

CHAPTER FOUR

4.71 The root-mean-square speed v_{rms} of a gas is $\sqrt{3RT/M}$. For two gases, the ratio is:

$$v_{H_2}/v_{O_2} = 1 = \sqrt{T_{H_2}/M_{H_2}}/\sqrt{T_{O_2}/M_{O_2}} = \sqrt{T_{H_2}M_{O_2}/T_{O_2}M_{H_2}}$$

$$\therefore T_{H_2} = T_{O_2}(M_{H_2}/M_{O_2}) = 298(2.016/32.00) = \mathbf{18.8K = -254.4°C}$$

4.73 a. Pressure is related to density and molar mass of an ideal gas by P = dRT/M, from which we can solve for M:
M = dRT/P = (8 g/L)(0.08206 L·atm/mol·K)(619K)/1.00 atm
= **4 × 10² g/mol**

b. The metal chloride contains 53.6% Cl and so it must contain 46.4% of the metal X. Let the formula be represented by X_aCl_b. Then the ratio of masses must be
M_X/M_{Cl} = 46.4 g/53.6 g = $(aM_X)/(bM_{Cl})$
0.866 = (a/b)M_X/35.5
M_X = 0.866(35.5)b/a = 30.7(b/a)
Since M_X is approximately 180, b/a must be 6/1. Therefore the exact molar mass is
M_X = 6(30.7) = **184 g/mol**

c. The metal is tungsten (M = 183.85 g/mol), and the formula of the chloride is **WCl₆**.

4.75 First determine the average molar mass of air:
M = $X_1M_1 + X_2M_2$ + ... where X is the fraction of each gas.
M = (0.781)(28.01 g/mol) + (0.209)(32.00 g/mol) + 0.010(39.95 g/mol)
= 29.0 g/mol
The density of an ideal gas is d = PM/RT.
d = (1.00 atm)(29.0 g/mol)/[(0.08206 L·atm/mol·K)(273K)]
= **1.29 g/L**

4.77 First write the balanced equation for the overall reaction:
$2S + 3O_2 + 2H_2O \rightarrow 2H_2SO_4$
From this, we see that the formation of 2 mol of H_2SO_4 requires 3 mol of O_2. Therefore, the number of mols of O_2 is
1000. kg H_2SO_4(10³g/1 kg)(1 mol H_2SO_4/98.07 g H_2SO_4)(3 mol O_2/2 mol H_2SO_4)
= 1.330 × 10⁴ mol O_2 required.

Air contains approximately 20.9 mol percent oxygen, so that
1.330 × 10⁴ mol O_2 × (1 mol air/0.209 mol O_2) = 7.318 × 10⁴ mol air required.

Now use the ideal gas equation to calculate the volume of air:
V = nRT/P = (7.318 × 10⁴ mol)(0.08206 L·atm/mol·K)(773K)/1.10 atm
= **4.22 × 10⁶ L** air required.

CHAPTER FOUR

4.79 We can calculate the total moles of CO and acetone that are present *initially*:
n = PV/RT = (0.132 atm)(1.0 L)/[(0.08206 L·atm/mol·K)(298K)]
n = 5.38×10^{-3} mol = $n°_{CO}$ + $n°_{acetone}$ (n° = initial moles).

After *complete reaction*, we can calculate the total moles of gas:
n = PV/RT = (0.150 atm)(1.0 L)/[0.08206 L·atm/mol·K)(298K)]
n = 6.13×10^{-3} = $n°_{CO}$ + $n_{products}$

But we know that *three* moles of gaseous products are produced for every mole of acetone consumed.

So $n_{products}$ = 3 $n°_{acetone}$. Substituting from above:
6.13×10^{-3} mol = $n°_{CO}$ + 3(5.38×10^{-3} mol − n_{CO})
2 $n°_{CO}$ = 0.0100 mol; $n°_{CO}$ = 5.00×10^{-3} mol

Since we now know the initial number of moles of CO, we can calculate the initial partial pressure of CO from
$P°_{CO}$ = $n°_{CO}$RT/V:
$P°_{CO}$ = (5.00×10^{-3} mol)(0.08206 L·atm/mol·K)(298K)/(1.0 L)
$P°_{CO}$ = 0.12 atm = **93 torr**

Since the three products are formed in a 1:1:1 mole ratio, we know that the number of moles of CO produced in the reaction is
(1/3)$n_{products}$ = (1/3)(6.13×10^{-3} mol) = 2.04×10^{-3} mol CO produced.

Therefore the total number of moles of CO after reaction is 7.04×10^{-3} mol.

The final partial pressure of CO can be calculated from
P = nRT/V = (7.04×10^{-3} mol)(0.08206 L·atm/mol·K)(298K)/(1.0 L)
= 0.17 atm = **130 torr**

4.81 A gas meter measures gas volume. Since gas density (m/V) decreases with an increase in temperature, a given mass of natural gas will have a larger volume—and thus cost more—in the summer as compared to the winter. The volume of an ideal gas is directly proportional to the absolute (Kelvin) temperature: V = nRT/P. Since the ratio of volumes at two temperatures is V_2/V_1 = T_2/T_1, the cost will vary in the same manner. The required ratio is V_2/V_1 = $cost_2/cost_1$ = 294K/261K = 1.13. Thus, the price for a given mass at 70°F will be **13% higher** than the price at 10°F.

CHAPTER FOUR

4.83 In the combustion process, C will be oxidized to CO_2, H will be oxidized to H_2O and S will be oxidized to SO_2. For every 1.00 kg of this coal, we have 835 g C, 51 g H, 9 g S and 97 g O. Based on the three equations:

$$C + O_2 \rightarrow CO_2$$
$$4H + O_2 \rightarrow 2H_2O$$
$$S + O_2 \rightarrow SO_2$$

we calculate

(835 g C)(1 mol C/12.01 g C)(1 mol O_2/1 mol C) = 69.5 mol O_2
(51 g H)(1 mol H/1.008 g H)(1 mol O_2/4 mol H) = 12.6 mol O_2
(9 g S)(1 mol S/32.06 g S)(1 mol O_2/1 mol S) = 0.3 mol O_2

The total number of moles of O_2 required, then, is 82.4 mol. Already present in the coal is the equivalent of 3.0 mol O_2, according to (97 g oxygen)(1 mol O_2/32.00 g) = 3.0 mol. So we need 79.4 additional mol O_2 from the air. We know that air contains about 20.9 mole percent O_2. Thus

(79.4 mol O_2)(1 mol air/0.209 mol O_2) = 380. mol air required.

Now we can calculate the volume of air required from the ideal gas relationship $V = nRT/P$:

V = (380. mol)(0.08206 L·atm/mol·K)(308K)/(0.920 atm)
   = **1.04×10^4 L**

4.89 Temperature is not specified, so we will assume 300K. Begin by determining the number of moles of air:

n = PV/RT
P = (30 lb/in^2)(51.715 torr/lb/in^2) × 1 atm/760 torr = 2.04 atm
V = 1 ft^3 = (0.02832 m^3)(10^3 L/m^3) = 28.3 L
T = 300K

n = (2.04)(28.3 L)/[(0.08206 L·atm/mol·K)(300K)] = 2.32 moles
M_{air} = 29.0 g/mol (as calculated in 4.75).
∴ m_{air} = 2.32 mol × 29.0 g/1 mol = **67 g of air**

4.91 Density is related to pressure and molar mass by

d = PM/RT
  = (0.001 atm)(720.6 g/1 mol)/[(0.08206 L·atm/mol·K)(300K)]
  = **0.029 g/L**

$v_{rms} = \sqrt{3RT/M}$; speed will come out in m/s if all quantities are in SI units.

$v_{rms} = \sqrt{3(8.314 \text{ J/K·mol})(300K)/0.7206 \text{ kg/mol}}$ = **1.0×10^2 m/s**

CHAPTER FIVE

Atomic Structure

5.1 Use the Einstein relation:
$E = mc^2 = (0.0040 \text{ kg})(3.00 \times 10^8 \text{ m/s})^2$
$= 3.6 \times 10^{14} \text{ kg·m}^2/\text{s}^2 = \mathbf{3.6 \times 10^{14} \text{ J}}$.

5.3 Rate of mass loss = 1×10^{-20} g/s = m/t
The time required to lose 10^{-6} g at this rate is
$t = 10^{-6} \text{ g}/(1 \times 10^{-20} \text{ g/s}) = 10^{14} \text{ s} = 10^{14} \text{ s} \times 1 \text{ hr}/3600 \text{ s} = \mathbf{3 \times 10^{10} \text{ hr}}$

5.5 $E = h\nu = hc/\lambda$.
 a. $E = (6.626 \times 10^{-34} \text{ J·s})(3.0 \times 10^8 \text{ m/s})/(5 \times 10^{-11} \text{ m})$
 $\mathbf{E = 4 \times 10^{-15} \text{ J}}$
 b. $E = (6.626 \times 10^{-34} \text{ J·s})(3.0 \times 10^8 \text{ m/s})/(1 \times 10^{-6} \text{ m})$
 $\mathbf{E = 2 \times 10^{-19} \text{ J}}$
 c. $E = (6.626 \times 10^{-34} \text{ J·s})(1 \times 10^{20} \text{ s}^{-1})$
 $\mathbf{E = 7 \times 10^{-14} \text{ J}}$

5.7 Energy and wavelength are related by $E = hc/\lambda$
$E = (6.626 \times 10^{-34} \text{ J·s})(3.00 \times 10^8 \text{ m/s})/7.24 \times 10^{-8} \text{ m}$
$= \mathbf{2.74 \times 10^{-18} \text{ J}}$

5.9 $E = h\nu = (6.626 \times 10^{-34} \text{ J·s})(1.21 \times 10^{15} \text{ s}^{-1}) = 8.02 \times 10^{-19} \text{ J} = 5.00 \text{ eV}$.
Since one photon will dissociate one molecule, the dissociation energy of molecular oxygen is $\mathbf{8.02 \times 10^{-19} \text{ J or 5.00 eV/molecule}}$

5.11 $W = hc/\lambda$, so $\lambda = hc/W$
$\lambda = (6.626 \times 10^{-34} \text{ J·s})(3.00 \times 10^8 \text{ m/s})/(7.7 \times 10^{-19} \text{ J})$
$= 2.6 \times 10^{-7} \text{ m} = \mathbf{260 \text{ nm}}$

5.13 Use work functions from Table 5.2. The work function equals the minimum energy required to expel electrons. For silver, this minimum energy is 7.37×10^{-19} J. Then
$\lambda = hc/W = (6.626 \times 10^{-34} \text{ J·s})(3.00 \times 10^8 \text{ m/s})/7.37 \times 10^{-19} \text{ J}$
$= 2.70 \times 10^{-7} \text{ m} = \mathbf{270. \text{ nm}}$
For copper,
$\lambda = (6.626 \times 10^{-34} \text{ J·s})(3.00 \times 10^8 \text{ m/s})/7.18 \times 10^{-19} \text{ J}$
$= 2.77 \times 10^{-7} \text{ m} = \mathbf{277 \text{ nm}}$

CHAPTER FIVE

5.15 a. $E = hc/\lambda$

$E = (6.626 \times 10^{-34}$ J·s$)(3.00 \times 10^8$ m/s$)/(1.81 \times 10^{-7}$ m$) = \mathbf{1.10 \times 10^{-18}}$ **J**

b. $W = \mathbf{7.69 \times 10^{-19}}$ **J**

5.17 a. $h\nu = W$

$\nu = 7.33 \times 10^{-19}$ J$/6.62 \times 10^{-34}$ J·s $= \mathbf{1.1 \times 10^{15}}$ **s^{-1}**

b. From part (a), KE = 0 at the threshold frequency

5.19 First, we need to determine the mass of an oxygen molecule.

$(32.00$ g/mol$)(1$ mol$/6.022 \times 10^{23}) = 5.314 \times 10^{-23}$ g

$\lambda = h/mv$

$\lambda = (6.626 \times 10^{-27}$ erg·s$)/(5.314 \times 10^{-23}$ g$)(50,000.$ cm/s$)$

$\lambda = 2.493 \times 10^{-9}$ cm $= \mathbf{0.2493}$ **Å**

5.21 a. $\lambda = (6.626 \times 10^{-27}$ erg·s$)/(5.0$ g$)(25,000$ cm/s$)$

$\lambda = 5.3 \times 10^{-32}$ cm $= \mathbf{5.3 \times 10^{-24}}$ **Å**

b. $(600.$ lb$)(454$ g/lb$) = 2.72 \times 10^5$ g

$(40.0$ mi/hr$)(1.609$ km/hr$)(1 \times 10^5$ cm/km$)(1$ hr$/3600$ s$) = 1.79 \times 10^3$ cm/s

$\lambda = (6.626 \times 10^{-27}$ erg·s$)/(2.72 \times 10^5$ g$)(1.79 \times 10^3$ cm/s$)$

$\lambda = 1.36 \times 10^{-35}$ cm $= \mathbf{1.36 \times 10^{-27}}$ **Å**

5.23 $\lambda = (6.626 \times 10^{-27}$ erg·s$)/(9.1 \times 10^{-9}$ g$)(3.0 \times 10^5$ cm/s$)$

$= \mathbf{2.4 \times 10^{-24}}$ **cm**

This wavelength is many orders of magnitude smaller than atomic dimensions. For all practical purposes its wave nature is invisible to us; it will behave as an "ordinary" macroscopic particle.

5.25 $\lambda = h/mv$; $v = h/m\lambda$

$v = (6.626 \times 10^{-34}$ J·s$)/[(9.11 \times 10^{-31}$ kg$)(10^{-10}$ m$)]$

$= \mathbf{10^7}$ **m/s.**

5.27 Ground state Be has the electronic configuration $1s^2 2s^2$. The four electrons are described by the following sets of quantum numbers:

$n = 1, \ell = 0, m_\ell = 0, m_s = +1/2$

$n = 1, \ell = 0, m_\ell = 0, m_s = -1/2$

$n = 2, \ell = 0, m_\ell = 0, m_s = +1/2$

$n = 2, \ell = 0, m_\ell = 0, m_s = -1/2$

CHAPTER FIVE

5.29 a. $\ell = 0, 1, 2$ b. $m_\ell = -3, -2, -1, 0, 1, 2, 3$ c. $m_s = +1/2, -1/2$

5.31 H $1s^1$
He $1s^2$
Li $1s^2 2s^1$
Be $1s^2 2s^2$
B $1s^2 2s^2 2p^1$
C $1s^2 2s^2 2p^2$
N $1s^2 2s^2 2p^3$
O $1s^2 2s^2 2p^4$
F $1s^2 2s^2 2p^5$
Ne $1s^2 2s^2 2p^6$
Na $1s^2 2s^2 2p^6 3s^1$
Mg $1s^2 2s^2 2p^6 3s^2$
Al $1s^2 2s^2 2p^6 3s^2 3p^1$
Si $1s^2 2s^2 2p^6 3s^2 3p^2$
P $1s^2 2s^2 2p^6 3s^2 3p^3$
S $1s^2 2s^2 2p^6 3s^2 3p^4$
Cl $1s^2 2s^2 2p^6 3s^2 3p^5$
Ar $1s^2 2s^2 2p^6 3s^2 3p^6$
K $1s^2 2s^2 2p^6 3s^2 3p^6 4s^1$
Ca $1s^2 2s^2 2p^6 3s^2 3p^6 4s^2$
Sc $1s^2 2s^2 2p^6 3s^2 3p^6 4s^2 3d^1$

5.33 a. As: [Ar]$4s^2 3d^{10} 4p^3$
b. Ti [Ar]$4s^2 3d^2$
c. I: [Kr]$5s^2 4d^{10} 5p^5$

5.35 The ten 4d electrons (n = 4, $\ell = 2$) have $m_\ell = -2, -1, 0, 1, 2$ and $m_s = \pm 1/2$.

$n = 4, \ell = 2, m_\ell = 2, m_s = +1/2$
$n = 4, \ell = 2, m_\ell = 2, m_s = -1/2$
$n = 4, \ell = 2, m_\ell = 1, m_s = +1/2$
$n = 4, \ell = 2, m_\ell = 1, m_s = -1/2$
$n = 4, \ell = 2, m_\ell = 0, m_s = +1/2$
$n = 4, \ell = 2, m_\ell = 0, m_s = -1/2$
$n = 4, \ell = 2, m_\ell = -1, m_s = +1/2$
$n = 4, \ell = 2, m_\ell = -1, m_s = -1/2$
$n = 4, \ell = 2, m_\ell = -2, m_s = +1/2$
$n = 4, \ell = 2, m_\ell = -2, m_s = -1/2$

CHAPTER FIVE

5.37   a. Rhodium (Rh)   b. Boron (B)

5.39   This element has the electron configuration [Ar]$4s^1$.
        The element is **potassium (K).**

5.41   a. 3d is higher than 4s.   b. 4p is higher than 4s.   c. 5s is higher than 4p.

5.43   From lower to higher energy, the orders are
        Na   1s, 2s, 2p, 3s
        K   1s, 2s, 2p, 3s, 3p, 4s

5.45   Na: [Ne]$3s^1$ K: [Ar]$4s^1$
        Ag: [Kr]$5s^1 4d^{10}$ Au: [Xe]$6s^1 5d^{10}$
        Removing the highest energy electron from Na or K will leave a noble gas electron configuration on the cation. Removing the highest energy electron from Ag or Au will disrupt a closed sub-shell of electrons.

5.47   a. Neutral, ground state   b. Anion, ground state
        c. Neutral, ground state   d. Impossible (no such thing as a 2d orbital)
        e. Anion, ground state

5.49   A ground state iron atom has a total of six d-electrons. According to Hund's rule, the electrons will go into empty, degenerate orbitals with spins parallel before they will fill each orbital with opposing spins. Iron has five d-electrons with parallel spins and one with an opposing spin.

5.51 $n_\ell = 4$ $n_u = 5$
        $1/\lambda = R_H(1/4^2 - 1/5^2) = 109{,}677.58$ cm^{-1} $(1/16 - 1/25) = 2.468 \times 10^3$ cm^{-1}
        $\lambda = 4.052 \times 10^{-4}$ cm $= 4.052 \times 10^{-6}$ m
        $E = hc/\lambda = (6.62 \times 10^{-34}$ J·s$)(3.00 \times 10^8$ m/s$)/4.052 \times 10^{-6}$ m
        $E = 4.90 \times 10^{-20}$ J

5.53   For sodium, $W = 3.67 \times 10^{-19}$ J.
        $KE = h\nu - W = hc/\lambda - W$
        $hc/\lambda = (6.626 \times 10^{-34}$ J·s$)(3.00 \times 10^8$ m/s$)/(4 \times 10^{-7}$ m$) = 4.97 \times 10^{-19}$ J
        $KE = (4.97 \times 10^{-19}$ J$) - (3.67 \times 10^{-19}$ J$) = 1.3 \times 10^{-19}$ J $= 1/2\, mv^2$
        1.3×10^{-19} J $= 1/2\,(9.1 \times 10^{-31}$ kg$)v_{max}^2$
        $v = 5 \times 10^5$ m/s

CHAPTER FIVE

5.55 The kinetic energy of an ejected photoelectron is $KE = h\nu - W = hc/\lambda - W$ where W is the work function of the metal. We are told that λ_1 is twice as long as λ_2. Since $E = hc/\lambda$, λ_1 has half the energy of λ_2. Let's designate the energy associated with λ_1 as E; then the energy of λ_2 is 2E. Then we can write

for λ_1: 1.6×10^{-19} J $= E - W$ and
for λ_2: $6.4 \times 10^{-19} = 2E - W$.

Subtracting the first equation from the second yields 4.8×10^{-19} J $= E$. Put this value into the equation for λ_1:

1.6×10^{-19} J $= 4.8 \times 10^{-19}$ J $- W$

$W = 3.2 \times 10^{-19}$ J.

5.57 a. For the outermost electron in ground state lithium, **n = 2, ℓ = 0**.
      b. For the highest energy electron in the first excited state of lithium, **n = 2, ℓ = 1**.
      c. For ground state Na, **n = 3, ℓ = 0**.
      In the first excited state, **n = 3, ℓ = 1**.

5.59 Cr: [Ar] $4s^1 3d^5$; [Ar] ↑(4s) ↑↑↑↑↑(3d)

Mn: [Ar] $4s^2 3d^5$; [Ar] ↑↓(4s) ↑↑↑↑↑(3d)

Cu: [Ar] $4s^1 3d^{10}$; [Ar] ↑(4s) ↑↓↑↓↑↓↑↓↑↓(3d)

Zn: [Ar] $4s^2 3d^{10}$; [Ar] ↑↓(4s) ↑↓↑↓↑↓↑↓↑↓(3d)

5.61 a. s-orbitals are non-degenerate (or "one-fold degenerate")
      b. p-orbitals are three-fold degenerate
      c. d-orbitals are five-fold degenerate.
      d. Se^{2-}: [Kr]   Cs: [Xe]$6s^1$

CHAPTER FIVE

5.63 The average translational kinetic energy of an ideal gas molecule at room temperature is KE = (3/2)kT.
KE = (1.5)(1.3806 × 10^{-23} J/K)(298K) = 6.17 × 10^{-21} J

a. ν = E/h = (6.17 × 10^{-21} J)/(6.626 × 10^{-34} J·s) = **9.31 × 10^{12} s^{-1}**

b. λ = c/ν = (3.00 × 10^8 m/s)/(9.31 × 10^{12} s^{-1}) = 3.22 × 10^{-5} m = **3.22 × 10^5 Å**

c. This wavelength of light is in the **radiowave** region of the electromagnetic spectrum.

d. Gas molecules at room temperature have sufficient kinetic energy to excite electrons from the ground state via collisions. We would not expect all atoms to be in their ground state at room temperature.

5.65 First, we will calculate the energy of the 340.2 nm line.
E = hc/λ (6.626 × 10^{-34} J·s)(2.998 × 10^8 m/s)/ (3.402 × 10^{-7} m)
E = 5.839 × 10^{-19} J

Next, we will calculate the energy of the 259.7 nm line.
E = hc/λ = (6.626 × 10^{-34} J·s)(2.998 × 10^8 m/s)/(2.597 × 10^{-7} m)
   = 7.649 × 10^{-19} J

The energy spacing is the difference between the two energies.
ΔE = (7.649 × 10^{-19} J) – (5.839 × 10^{-19} J)
ΔE = (1.810 × 10^{-19} J)(1 kJ/10^3 J)(6.022 × 10^{23}/mol)
ΔE = 109.0 kJ/mol

5.67 First, we will calculate the energy per photon from the equation E = hc/λ.
E = (6.626 × 10^{-34} J·s)(3.00 × 10^8 m/s)/(6.60 × 10^{-7} m) = 3.01 × 10^{-19} J/photon
In each second, 5 × 10^{19} photons are emitted. This corresponds to a total energy of
(5 × 10^{19})(3.01 × 10^{-19} J) = 15 J.

Power = (15 J)/(1 s) = **15 watts**

The energy per photon can be converted into eV and compared with the other examples given.
(3.01 × 10^{-19} J)(1eV/1.6022 × 10^{-19} J) = 1.9 eV. One photon of 660-nm light is

a. more energy than a water molecule falling over Niagara Falls
b. comparable to the energy released when one carbon atom is burned to formed carbon dioxide.
c. much less than the energy released by the fission of one U^{235} atom

CHAPTER FIVE

5.71 Planck's constant has the units Joule·seconds. A Joule is a Newton·meter, and a Newton is the force that, when acting on a mass of 1 kg, produces an acceleration of 1 m/s^2. Therefore, a Joule is 1 kg·m^2/s^2. Multiplying this by seconds gives the unit of Planck's constant: kg·m^2/s.

Angular momentum is the product of mass, velocity and radius: mvr. As a consequence, its units are (kg)(m/s)(m) = kg·m^2/s. Thus, these are the same units as those of Planck's constant.

In his model of the atom, Bohr allowed only values of angular momentum that were integral multiples of h/2π: mvr = nh/2π, where n = 1, 2, 3...= the principal quantum number.

5.73 First, determine the energy associated with the 320 nm wavelength:
E = hc/λ = (6.626 × 10^{-34} J·s)(3.00 × 10^8 m/s)/(320 × 10^{-9} m) = 6.2 × 10^{-19} J
Any metal with a work function greater than 6.2 × 10^{-19} J will not eject electrons if irradiated with radiation having a wavelength less than 320 nm light. From Table 5.2, we see this would include **zinc, copper, gold, platinum and silver**.

5.75 1/λ = R$_H$(1/n$_1^2$ − 1/n$_2^2$)
1/λ = (109,678 cm^{-1})(1/49^2 − 1/50^2) = 1.809 cm^{-1}
λ = 0.553 cm = **5.53 mm**

This wavelength of light is in the **radiowave** region of the electromagnetic spectrum.

CHAPTER SIX

Properties of Molecules

6.1 a. $[:\ddot{\underset{..}{O}}-H]^{\ominus}$ b. $H-\overset{..}{\underset{H}{O}}-H$ with $\oplus$ charge c. $H-\overset{\overset{H}{|}}{\underset{\underset{H}{|}}{N}}-H$ with $\oplus$ charge

d. $[H-\overset{..}{\underset{..}{N}}-H]^{\ominus}$ e. $[:\overset{..}{\underset{..}{S}}-H]^{\ominus}$ f. $H-\overset{..}{\underset{H}{S}}-H$ with $\oplus$ charge

6.3 a. $:\overset{\overset{..}{\underset{|}{F}:}}{\underset{\underset{..}{:F:}}{\overset{..}{F}}}-\overset{..}{\underset{..}{B}}-\overset{..}{\underset{..}{F}}:$ with $\ominus$ charge b. $H-\overset{..}{\underset{..}{O}}-\overset{..}{\underset{..}{O}}:$ with $\ominus$ charge c. $H-O-\overset{\overset{:\overset{..}{O}:}{|}}{\underset{\underset{:\overset{..}{O}:}{|}}{S}}-\overset{..}{\underset{..}{O}}:$ with $\ominus$ charge

6.5 a. $H-\overset{..}{\underset{..}{O}}-H$ b. $H-\overset{..}{\underset{..}{O}}-\overset{..}{\underset{..}{O}}-H$ c. $\overset{..}{\underset{..}{O}}=C=\overset{..}{\underset{..}{O}}$ d. $\overset{\overset{:\overset{..}{O}:}{\|}}{\underset{\underset{H\;\;\;\;\;\;\;H}{\overset{..}{\underset{..}{O}}\;\;\;\;\overset{..}{\underset{..}{O}}}}{C}}$

e. $H-\overset{\overset{H}{|}}{\underset{\underset{H}{|}}{C}}-\overset{..}{\underset{..}{O}}-H$ f. $H-\overset{\overset{H}{|}}{\underset{\underset{H}{|}}{C}}-\overset{\overset{H}{|}}{\underset{\underset{H}{|}}{C}}-\overset{..}{\underset{..}{O}}-H$ or $H-\overset{\overset{H}{|}}{\underset{\underset{H}{|}}{C}}-\overset{..}{\underset{..}{O}}-\overset{\overset{H}{|}}{\underset{\underset{H}{|}}{C}}-H$

6.7 a. $H-\overset{..}{\underset{..}{N}}=N=\overset{..}{\underset{..}{N}}$ b. $H-\overset{..}{\underset{\underset{H}{|}}{N}}-C\equiv N:$ c. $H-\overset{\overset{H}{|}}{\underset{\underset{H}{|}}{C}}-\overset{..}{\underset{\underset{H}{|}}{N}}-H$

6.9 $\left[:\overset{..}{\underset{..}{O}}-\overset{\overset{..}{\underset{\underset{:\overset{..}{O}:}{|}}{S}}}{}-\overset{..}{\underset{..}{O}}:\right]^{2-}$

6.11 Formal charge fc = no. valence electrons – no. unpaired electrons – 1/2 no. shared electrons.

a. 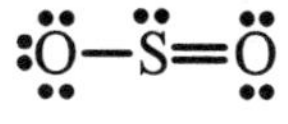 or 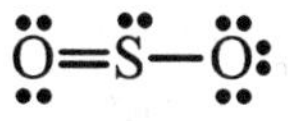

fc: -1 +1 0 0 +1 -1

Fc is calculated as follows for the first structure:

left O: fc = 6 – 6 – 2/2 = –1

S: fc = 6 – 2 – 6/2 = +1

right O: fc = 6 – 4 – 4/2 = 0

b. 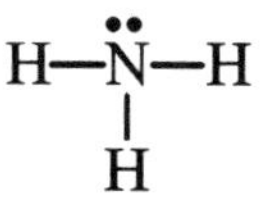

H: fc = 2 – 2/2 = 0

N fc = 5 – 2 – 6/2 = 0

c. 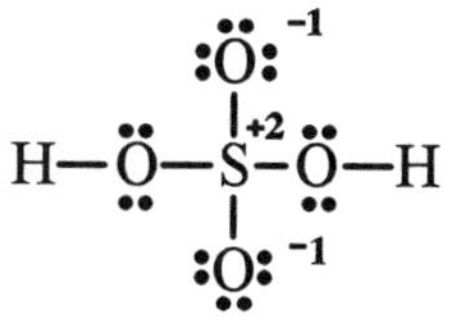

Formal charges are all zero except as indicated

S: fc = 6 – 0 – 8/2 = +2

each O: fc = 6 – 6 – 2/2 = –1

6.13 There are two resonance forms:

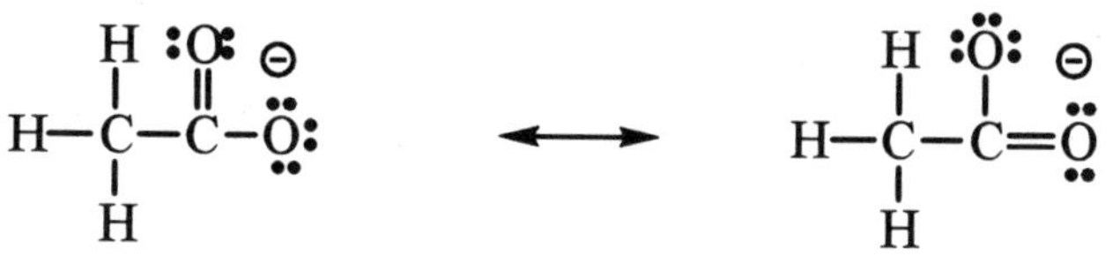

6.15 a. 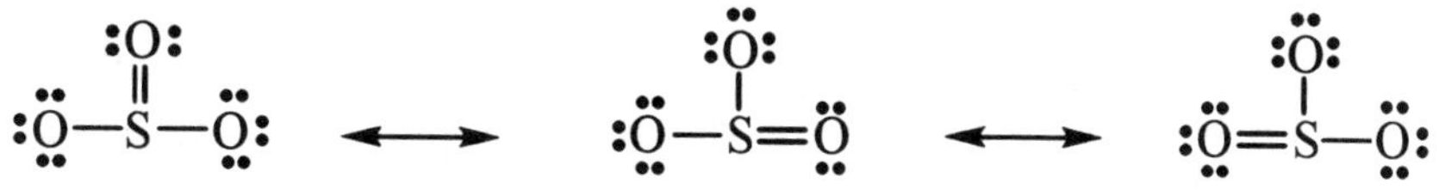

b. 

CHAPTER SIX

6.17 Ionic vs. covalent character of a bond can be predicted from the difference in electronegativity (EN) between the atoms. A rule of thumb is to assign roughly 50% ionic character to a bond with a difference of 1.5 - 2 units in EN.
   a. EN(Cs) = 0.7;   EN(F) = 4.0;   **ionic**
   b. EN(Ba) = 0.97;   EN(Br) = 2.8;   **ionic**
   c. EN(O) = 3.5;   EN(N) = 3.0;   **covalent**
   d. EN(H) = 2.1;   EN(Cl) = 3.0;   **covalent**

6.19 The ionic compounds are **CsF** and **BaBr$_2$**.

6.21 Atoms H and F have the largest electronegativity difference, so **HF** has the largest dipole moment.

6.23 a. O is positive; F is negative
   b. N is positive; O is negative
   c. S is positive; O is negative
   P-H, C-S and **N-Cl** are three covalent bonds between dissimilar elements whose electronegativities are nearly the same, so the bonds are essentially nonpolar.

6.25 Larger atoms with more electrons are more polarizable. The atom with the stronger induced dipole, and therefore higher boiling point, is **Xe**.

6.27 Molecules with more electrons will be more polarizable. Polarizability and therefore boiling points increase in the order **H$_2$ < N$_2$ < O$_2$ < F$_2$**.

6.29 The two Lewis structures are

$$\text{H}-\ddot{\text{O}}-\overset{\overset{\displaystyle :\ddot{\text{O}}:}{|}}{\text{N}}=\ddot{\text{O}} \quad \text{and} \quad \text{H}-\ddot{\text{O}}-\overset{\overset{\displaystyle :\text{O}:}{\|}}{\text{N}}-\ddot{\ddot{\text{O}}}:$$

They are equivalent.

6.31 Intermolecular forces can be either attractive or repulsive; dipole and induced dipole interactions result in attractive forces between gas particles. Gases which can be expected to be most nearly ideal are those having no permanent dipoles and only small induced dipoles. Examples are helium, neon, hydrogen, nitrogen, and oxygen. Conversely, gases with large permanent dipoles (e.g. H$_2$O, NH$_3$) or large induced dipoles (Cl$_2$, C$_3$H$_8$) will be the most non-ideal. These attractive forces cause the PV product to be smaller than ideal, and so their compressibility factors Z will typically be less than unity

CHAPTER SIX

6.33   a. First, sulfur dioxide reacts with oxygen in the air to form sulfur trioxide.
$$2SO_2 + O_2 \rightarrow 2SO_3$$
Then sulfur trioxide reacts with water to form sulfuric acid:
$$SO_3 + H_2O \rightarrow H_2SO_4$$

b. In one hour, 25 tons of copper are produced.
$(25 \times 10^6 \text{ g Cu})(1 \text{ mol}/63.55 \text{ g}) = 3.93 \times 10^5$ mol Cu
This copper was produced from 3.93×10^5 mol $CuFeS_2$ which also produces 7.87×10^5 mol SO_2.
$(7.87 \times 10^5 \text{ mol } SO_2)(64.06 \text{ g}/1 \text{ mol}) = 5.04 \times 10^7$ g SO_2 = **50. tons SO_2 per hour**

c. If the chalcopyrite is contaminated with iron pyrites, then more SO_2 will be produced than can be estimated from the output of copper. The estimate in part (b) will be **too low.**

6.39   The energy of a photon is $E = h\nu = hc/\lambda$
$E = (6.626 \times 10^{-34} \text{ J·s})(3.00 \times 10^8 \text{ m/s})/(4.800 \times 10^{-7} \text{ m})$
    $= 4.138 \times 10^{-19}$ J/photon $= (4.138 \times 10^{-19}$ J/photon$)(6.022 \times 10^{23}$ photons/mol$)$
    $= 249.4$ kJ.
The bond energy is the photon energy minus that of the excited chlorine atom:
BE = 249.4 − 10.5 = **239 kJ/mol.**

CHAPTER SEVEN

Theories of Chemical Bonding

7.1 First write the Lewis dot structures for the molecules and ions. BF$_3$ is unusual in that B does not achieve an octet (B is "electron deficient").

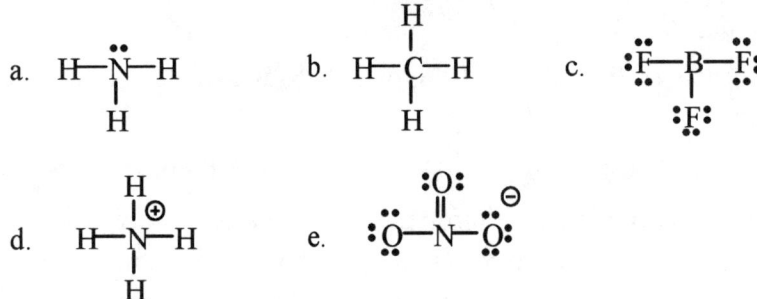

BF$_3$ and NO$_3^-$ have similar electron arrangements about the central atoms, and their shapes are both trigonal planar:

$$\begin{array}{c} \ddot{\ddot{F}} \\ | \\ \ddot{\ddot{F}}^{\,\,B}\ddot{\ddot{F}} \end{array} \quad \text{and} \quad \begin{array}{c} \ddot{\ddot{O}} \\ \| \\ \ddot{\ddot{O}}^{\,\,N}\ddot{\ddot{O}} \end{array}$$

The other molecules and ion have the same electron geometries: tetrahedral due to four electron pairs about the central atoms. The shapes of CH$_4$ and NH$_4^+$ are also tetrahedral, but that of NH$_3$ is pyramidal because of the nonbonded electron pair on nitrogen.

7.3 Both H$_2$S and H$_2$O have Lewis structures with 2 bonded electron pairs and 2 non-bonded pairs:

$$\begin{array}{c} \ddot{\ddot{S}}\text{—H} \\ | \\ H \end{array} \quad \begin{array}{c} \ddot{\ddot{O}}\text{—H} \\ | \\ H \end{array}$$

So **yes**, they should have similar shapes Both in fact are bent molecules.

38

CHAPTER SEVEN

7.5   a. **SCl₂** has two bonds to Cl, and two non-bonded pairs on S:

Its electron geometry is tetrahedral and the molecular geometry is **bent**, or angular. Similarly, **SF₄** has four bonds to F and one non-bonded pair on S; its electron geometry is trigonal bipyramidal and its molecular geometry is **see-saw shaped**. **Cl₂PF₃** has five bonds to Cl and F and no non-bonded pairs on P; electron and molecular geometries are both **trigonal bipyramidal**. **SbF₃** has three bonds to F and one non-bonded pair on Sb; its electron geometry is tetrahedral and its molecular geometry is **trigonal pyramidal**. **AsCl₅** is similar to Cl₂PF₃; its molecular geometry is **trigonal bipyramidal**. **BrF₂⁻** has two bonds to F and three non-bonded pairs on Br. Its electron geometry is trigonal bipyramidal and its molecular geometry is **linear**.

    b. SCl₂ is sp³   SF₄ is sp³d
       Cl₂PF₃ is sp³d   SbF₃ is sp³
       AsCl₅ is sp³d   BrF₂⁻ is sp³d

7.7

	a.	b.
KrF₂	2 bonds and 3 non-bonded pairs on Kr: **linear**	sp³d
XeF₄	4 bonds and 2 non-bonded pairs on Xe: **square planar**	sp³d²
BrF₃	3 bonds and 2 non-bonded pairs on Br: **T-shaped**	sp³d
IF₅	5 bonds and 1 non-bonded pair on I: **square pyramidal**	sp³d²
BeI₂	2 bonds and no non-bonded pairs on Be: **linear**	sp

7.9 

                                                    plus one more resonance form

7.11   a. HC≡C-CH=CH₂ contains 7 σ-bonds and 3 π-bonds.
       b. CH₂=CH-COOH contains 8 σ-bonds and 2 π-bonds.

CHAPTER SEVEN

7.13 Each Be has two 2s electrons. There are two σ_{2s} bonding electrons and two $(\sigma_{2s})^*$ antibonding electrons. The bond order is zero so Be_2 doesn't exist.

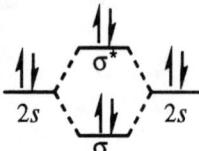

7.15 The **valence bond view** explains the formation of O_2 as follows:

Each oxygen atom has 6 valence electrons, two short of an octet. Also, each O atom has two unpaired electrons in p-orbitals, available for sharing. So each O needs two more electrons, and each has two to share. A total of four electrons are shared (two from each O). A double bond results.

The **molecular orbital view** explains the formation of O_2 as follows: Each oxygen atom's 2s and three 2p orbitals form molecular orbitals. The 2s orbitals combine to form a σ (bonding) and a σ^* (antibonding) molecular orbital. The 2 p orbitals combine to form, σ, σ^*, two π and two π^* molecular orbitals. The ten valence electrons fill the σ_{2s}, σ_{2s}^*, σ_{2px}, π_{2py} and π_{2pz} molecular orbitals. There are eight bonding electrons and four antibonding electrons for a net bond order of 2. A double bond results.

7.17 O_2^+ has eleven valence electrons.
$KK(\sigma_{2s})^2 (\sigma^*_{2s})^2 (\pi_{2p})^4 (\sigma_{2p})^2 (\pi^*_{2s})^1$

7.19 **He_2^+ and O_2 are paramagnetic.**
H_2, Li_2, Be_2 (if it existed) and N_2 are all diamagnetic.

7.21 No, the orientation of the p_x and p_y orbitals is such that the distance between them is too great and there is insufficient overlap to form a good bond.

7.23 a. The species does exist.   b. The apparent bond order is 1.
c. It is diamagnetic.   d. The bond length is shorter than in Li_2^+.
e. The bond length is longer than in H_2.
f. The energy released is greater than in Na_2 formation.
g. The energy released is less than in B_2 formation.

CHAPTER SEVEN

7.25 The tri-iodide ion, I_3^-, has 22 valence electrons. The valence bond structure will be an expanded shell about the central iodine:

$$I-\overset{..}{\underset{..}{I}}-I$$

where the non-bonding electron pairs on the terminal iodines are not shown. According to the VSEPR model, this ion is linear.

7.27 P_4 has 20 valence electrons. Possible structures include a square or tetrahedral arrangement of atoms:

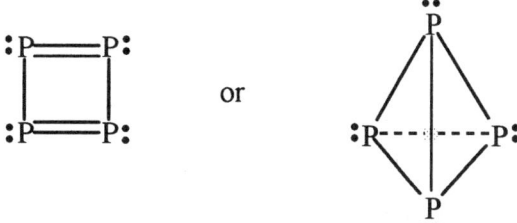

7.29 All the individual bonds in these molecules are polar. Whether the molecules themselves are polar depends on whether they have net dipoles resulting from unsymmetrical geometries.

The following molecules are not symmetrical and therefore are the most polar: **NH_3** (pyramidal) and **ClF_3** (T-shaped)

The remaining molecules are completely symmetrical and have no net dipoles: **BCl_3** (trigonal), **PCl_3** (trigonal bipyramidal) and **XeF_4** (square planar)

7.31 Of the molecules given, BCl_3 is electron deficient: there are only six electrons about boron. It can form a co-ordinate covalent bond by accepting (sharing) an electron pair from a molecule having a non-bonding pair. An example is

$$\underset{Cl}{\overset{Cl}{Cl-B}} \;+\; :NH_3 \longrightarrow Cl_3B-NH_3$$

7.33 The molecule and ion both have ten valence electrons. They have similar structures:

$$:P\equiv P: \quad \text{and} \quad :C\equiv C:^{2-}$$

with a triple bond and two non-bonding pairs. They each have one σ bond and two π bonds, and are sp hybridized.

CHAPTER SEVEN

7.35 SF$_6$ is symmetrical, non-polar, and has no non-bonded electron pairs on S. Conversely, SF$_4$ is more reactive because of its lone pair of electrons and molecular polarity. SF$_4$ is particularly reactive with polar molecules such as H$_2$O because of their mutual polar attraction.

7.37 :N≡C–Ö:$^\ominus$ is the most favorable resonance form. Others that could be written are

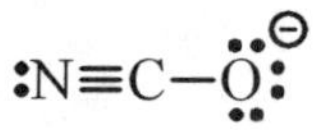

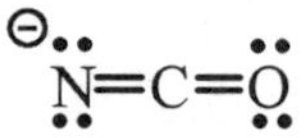

$^\ominus$:N̈=C=Ö: and $^{2-}$:N̈–C≡O:$^\oplus$

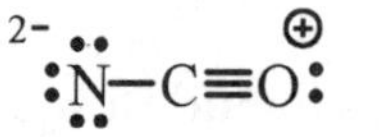

7.43 Ozone, with $3 \times 6 = 18$ valence electrons, has the two resonance forms:

We will assume that each oxygen atom is sp^2 hybridized. That is, the s orbitals and two of the three p orbitals combine to form 3 orbitals of identical symmetry. This leaves one p orbital that is not hybridized and will act as a normal p orbital. First, we will consider bonding in the plane of the molecule only:

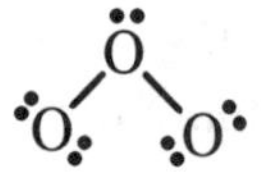

All of the bonding in the plane of the molecule *via* sp^2 orbitals, which are arranged around each atom like points on a triangle. The two terminal oxygen atoms each have two sp^2 orbitals occupied with lone pairs, and the central atom has one sp^2 orbital with a lone pair. The lone pair electrons are not involved in bonding and will not be included in the molecular orbitals.

Each terminal oxygen atom has one sp^2 orbital containing one electron. The central atom has two sp^2 orbitals each filled with one electron. These orbitals form σ bonds with the half-filled sp^2 orbitals on the terminal atoms.

The 5 lone pairs and two σ bonds account for 14 of the 18 valence electrons. The 4 remaining electrons form π molecular orbitals. Because of sp^2 hybridization, their p orbitals are perpendicular to the plane of the molecule (imagine them sticking out of and below the plane of the page in the diagram above) and so are properly oriented for π bonding. Since each participating atomic orbital (AO) will make one molecular orbital (MO), we will have:

Central atom:	2 sp² orbitals involved in σ bonding
	1 p orbital involved in π bonding
	1 lone pair, not involved in bonding
Terminal atoms:	2 sp² orbitals involved in σ bonding
	2 p orbitals involved in π bonding
	4 lone pairs, not involved in bonding

Here is the resulting molecular orbital diagram for the three oxygen atoms in ozone:

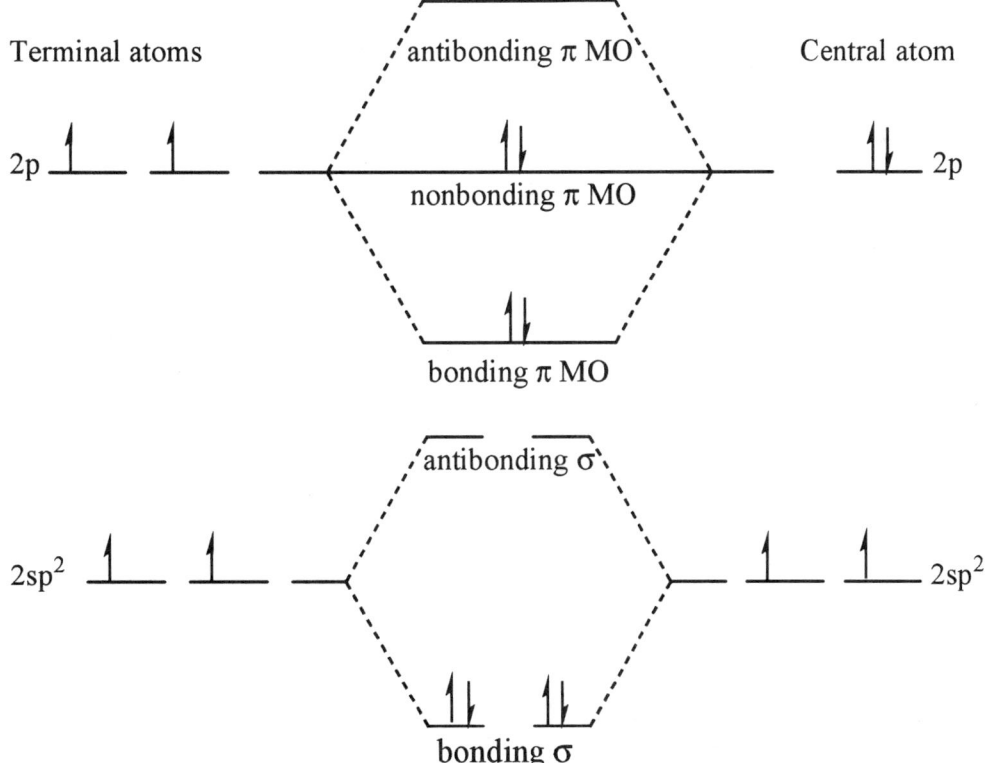

CHAPTER SEVEN

7.45 Each intersection represents a carbon atom, and sp² means that all carbon atoms have 3 bonding centers. Draw alternating single and double bonds throughout, and use H as needed so that there are 8 electrons on each carbon.

The compound is $C_{20}H_{12}$.

CHAPTER EIGHT

Periodic Properties

8.1   a. Ge is in the **Group IVA**.
b. Sr is in **Group IIA**, the alkaline earth elements.

8.3 I_1 tends to increase toward the top right corner of the periodic table. However, there is a small decrease in I_1 at the next element after filling or half filling a set of orbitals.
a. **K < Mg < Be**   b. **Na < Al < Mg**

8.5   Electron affinities increase up and to the right of the Periodic Table. Therefore, the order is **K < Li < I < Cl**.

8.7   In general, combinations of metals with nonmetals will yield ionic salts.
BaO, KF and **SrO** will be ionic salts. CO, ClF, NO and HI will be covalent molecules.

8.9   a. H is +1, and N is −3.
b. K is +1, O is -2 and Mn is +7.
c. O is −2, and Br is +5.
d. Ca is +2, O is −2 and C is +4.
e. Na is +1, F is −1 and Cl is +3.

8.11 a. $Cs + 1/2\ Br_2 \rightarrow CsBr$
b. $Na + 1/2\ H_2 \rightarrow NaH$

8.13 a. $C_3H_8 + 5O_2 \rightarrow 3CO_2 + 4H_2O$
b. $N_2H_4 + O_2 \rightarrow N_2 + 2H_2O$

8.15 Here are the balanced equations for the reactions:
a. $Ba + H_2 \rightarrow \mathbf{BaH_2}$
b. $Ba + 2H_2O \rightarrow \mathbf{Ba(OH)_2 + H_2}$
c. $Ba + I_2 \rightarrow \mathbf{BaI_2}$

8.17 a. $Al^{3+} < Mg^{2+} < Na^+$
b. $F^- < Cl^- < Br^-$
c. $Mg^{2+} < Na^+ < Cl^-$

8.19 a. Li < Na < Rb
b. F < N < B
c. Cl < P < Ga

CHAPTER EIGHT

8.21 From Table 6.1a, $r_O = 0.73$ Å, $r_F = 0.72$ Å
∴ d_{OF} = **1.45 Å**

8.23 Atomic radii increase going down and left in the periodic table:
Al < Ca < Sr < Rb < Cs

8.25 Except for the very largest cations, cations are smaller than any of the anions. Within each group, size increases to the left of the periodic table and down a column
$Al^{3+} < Mg^{2+} < Ca^{2+} < O^{2-} < Cl^- < S^{2-}$

8.27 Nitrogen monoxide catalyzes the reaction $O_3 + O \rightarrow 2O_2$ by the following steps:
$NO + O_3 \rightarrow NO_2 + O_2$
$NO_2 + O \rightarrow NO + O_2$ etc...

8.29 $4Li + O_2 \rightarrow 2Li_2O$
$2Na + O_2 \rightarrow Na_2O_2$
$K + O_2 \rightarrow KO_2$
$Rb + O_2 \rightarrow RbO_2$
$Cs + O_2 \rightarrow CsO_2$
Other reactions are also possible.

8.31 $CaCO_3 \rightarrow CaO + CO_2$
$CaO + H_2O \rightarrow Ca(OH)_2$

8.33   a. NaH is an **ionic** hydride. It is a **solid** at room temperature.
  b. HF is a **covalent** compound. It is a **gas** at room temperature. However, it is reasonably easy to liquefy due to extensive hydrogen-bonding between HF molecules.
  c. HCl is a **covalent** compound. It is a **gas** at room temperature. It is not as easy to liquefy as HF because Cl is less electronegative than F and the hydrogen-bonding is not as strong.
  d. SiH_4 is a **covalent** compound. It is a **gas** at room temperature.
  e. SnH_4 is a **covalent** compound. It is a **gas** at room temperature.
  f. NH_3 is a **covalent** compound. It is a **gas** at room temperature.
  g. B_2H_6 is a **covalent** compound. It is a **gas** at room temperature.
  h. Na_2S is an **ionic** compound. It is a **solid** at room temperature.
  i. H_2S is a **covalent** compound. It is a **gas** at room temperature

CHAPTER EIGHT

8.35 $Pb + O_2(excess) \rightarrow PbO_2$
$PbO_2 + 4HCl \rightarrow PbCl_4 + 2H_2O$

8.37 The reaction is $Na(g) \rightarrow Na^+(g) + e^-$.
This is simply the ionization energy of Na(g), which from Table 8.2 is **496 kJ/mol**.

8.39 Acids can be defined as substances that lose H^+ to form the H_3O^+ ion in water. The fact that H_2S is more strongly acidic than H_2O means that the **H-S bond is weaker** than the H-O bond.

8.41 Write the electron configurations using normal notation:
Z = 4 is Be: $[He]2s^2$ s-block
Z = 8 is O: $[He]2s^22p^4$ p-block
Z = 17 is Cl: $[Ne]3s^23p^5$ p-block
Z = 19 is K: $[Ar]4s^1$ s-block
Z = 22 is Ti: $[Ar]4s^23d^2$ d-block

8.43 We will start by writing a balanced equation for the reaction.
$4HF + SiO_2 \rightarrow SiF_4 + 2H_2O$
Now find the number of moles of HF.
(4.00 L)(0.750 mol/L) = 3.00 mol HF
3.00 mol HF can produce 0.750 mol SiF_4, but since the yield is only 85.0%, we will produce only (0.850)(0.750 mol) = 0.638 mol SiF_4.
Now calculate the volume that 0.638 mol SiF_4 will occupy under the given conditions, assuming ideal behavior:
V = nRT/P = (0.638 mol)(0.08206 L·atm/mol·K)(298K)/(1.00 atm)
V = 15.6 L SiF_4

8.49 Trends in melting points:
a. K < Na < Mg < Al < Si
Valence electrons are held more tightly proceeding from left to right.
b. Ar < Cl < P < S
Argon is chemically inert and therefore does not have any bonding interactions to increase its mass and melting point. Chlorine forms Cl_2 molecules, phosphorus P_4 molecules and sulfur S_8 rings, all having increasing molar mass in the order given.

CHAPTER EIGHT

8.51 From the Bohr model in Chapter 5, we know that the energy associated with an atom of nuclear charge Z containing a single electron is $E_n = -B(1/n^2)$ where $B = (2\pi^2 me^4 Z^2)/h^2$ and the energy difference between any two states is $E_2 - E_1 = -B(1/n_2^2 - 1/n_1^2) = +B(1/n_1^2 - 1/n_2^2)$. Complete removal of an electron from the atom corresponds to n_2 = infinity, and the energy required for this process starting from the n = 1 level is the ionization energy:
$IE = E_\infty - E_1 = B/n_1^2 = (2\pi^2 me^4 Z^2)/[(n_1^2)h^2]$

We are given that the ionization energy is 5250 kJ/mol. For convenience, we will convert this to ergs per atom:
(5250 kJ/mol)(10^{10} kJ/erg)/(6.022 × 10^{23} atoms/mol) = 8.72 × 10^{-11} erg/atom.

Now solve for the nuclear charge Z from the known values of electron mass and charge:
(8.72 × 10^{-11} erg/atom)
$= 2\pi^2(9.11 \times 10^{-28} g)(4.803 \times 10^{-10} esu)^4 Z^2/[(1^2)(6.626 \times 10^{-27} erg \cdot s)^2]$

Solving gives $Z^2 = 4$, and therefore **Z = 2**.

8.53 $ClCF_3$ has a molar mass of 104.459 g/mol and contains 0% Br. $HBrCF_2$ has a molar mass of 112.921 g/mol. It contains 79.904 g/mol Br/112.921 g/mol × 100 = 70.76 % Br

Let x = fraction of $HBrCF_2$ in the mixture of the two gases. This is related to the percent of bromine in the mixture, 24.0%, by
24.0% = x(70.76%) + (1 – x)(0%). Solve: x = 0.339

The mixture is **33.9% $HBrCF_2$ and 66.1% $ClCF_3$**.

Neither of the two gases is environmentally attractive, but $HBrCF_2$ has one hydrogen attached to a carbon and a relatively weak C-Br bond, so it will decompose more easily than will $ClCF_3$. Both compounds eventually will decompose to other halogen-containing compounds, which may find their way into the ecosystem of the earth.

8.55 In the stratosphere, the following reactions occur:
O_3 + ultraviolet radiation → O_2 + O·
$2O_2$ + ultraviolet radiation → O_3 + O·

Both processes absorb radiation, increasing the temperature of the stratosphere. The temperature rises with altitude because there is more UV radiation at higher altitudes. The process of ozone, oxygen and oxygen atom formation is self-sustaining left to itself.

CHAPTER NINE

Liquids and Solutions

9.1 The normal boiling point of any substance is defined as the temperature at which its vapor pressure is one atm. Therefore, the vapor pressure of aluminum at 2450°C is **one atm.**

9.3 Plot vapor pressure vs. temperature for CCl_4. The temperature required to achieve a vapor pressure of 760 torr is 77°C. **77°C** is the normal boiling point of CCl_4.

9.5 The normal boiling point should fall somewhere near **100°C** in each case. However, the $\log(p_{vap})$ vs. $10^3/T$ plot should be a nearly straight line and should give a more accurate value of the boiling point.

9.7 Vapor pressure data for water is provided in Table 9.2. We know that boiling occurs when vapor pressure is equal to the pressure of the atmosphere. Therefore, water will boil in Denver when its vapor pressure is equal to 630 torr and water will boil in Rocky Mountain National Park when its vapor pressure is equal to 550 torr. Plot vapor pressure vs. temperature for water. You should find that the boiling points of water at these two locations are approximately **97°C** and **91°C,** respectively.

9.9 The triple point is the condition of temperature and pressure at which the three phases solid, liquid, and vapor are in equilibrium. For water (see Fig. 9.11), the triple point is at +0.01°C and 4.6 torr.

9.11 a. The slope of the solid-liquid equilibrium line is positive. This means that the solid phase is more dense than the liquid phase since an increase in pressure will favor formation of the solid phase. An increase in pressure then will make melting less favorable. Increasing the pressure will increase the melting point.
 b. At 373K and 1.25 atm, this substance is a liquid. Decreasing the pressure at constant temperature (a vertical line on the graph) will eventually cause the liquid to boil and exist as a vapor. Decreasing the temperature at a constant pressure (a horizontal line on the graph) will eventually cause the liquid to freeze and exist as a solid.
 c. The normal melting point of this substance is approximately **345K** and its normal boiling point is approximately **390K**. At room temperature (*ca.* 300K), the substance is a solid. The solid's vapor pressure (or sublimation pressure) is approximately **0.25 atm** at 300K.

CHAPTER NINE

9.13 a. The critical pressure is the pressure required to liquefy a gas at its critical temperature. Above the critical temperature, a gas cannot be liquefied no matter how much pressure is applied. The critical point (temperature and pressure) occurs at the point where the liquid-vapor equilibrium line comes to an end. For substance R, the critical pressure is approximately **75 atm**.
b. The slope of the solid-liquid equilibrium line is positive. This means that the solid phase is more dense than the liquid phase since an increase in pressure will favor formation of the solid phase. Therefore, solid R **will sink** in its own liquid.
c. The triple point of R is **8°C** and **25 atm**. This is the point at which solid, liquid and vapor coexist in equilibrium in the same container.
d. This substance does not have a normal boiling point. At 1 atm pressure, solid R undergoes sublimation. The liquid phase does not exist at this pressure so there can be no boiling point at 1 atm.

9.15 Molarity is the number of moles of solute per liter of solution: molarity = n/V. Therefore, n_{CH_3OH} = (1.50 moles/L)(0.425 L) = **0.638 moles**

9.17 Since we know that the masses are equal, let us assume that we have one gram of each. Then we can convert the masses into moles and calculate the mole fractions.
(1.00 g C_2H_5OH)(1 mol/46.07 g) = 0.0217 mol C_2H_5OH
(1.00 g H_2O)(1 mol/18.02 g) = 0.0555 mol H_2O
(1.00 g $C_2H_4(OH)_2$)(1 mol/62.07 g) = 0.0161 mol $C_2H_4(OH)_2$
The total number of moles then is n_{tot} = 0.0933 mol.
We can now calculate the mole fractions by the relationship $X_A = n_A/n_{tot}$.
$X_{ethanol}$ = (0.0217/0.0933) = **0.232**
X_{water} = (0.0555/0.0933) = **0.595**
$X_{ethylene\ glycol}$ = (0.0161/0.0933) = **0.172**

9.19 In order to calculate the molality of each of these solutions, we need to know the number of moles of solute and the number of kilograms of solvent. Then we can make use of the relationship molality = moles solute/kilograms solvent.
a. (40. g NaOH)(1 mol/40.00 g) = 1.0 mol NaOH
(1.0 mol NaOH)/(0.400 kg H_2O) = **2.5 m NaOH**
b. (0.032 g CH_3OH)(1 mol/32.04 g) = 0.0010 mol CH_3OH
(10. mL H_2O)(1.0 g/mL)(1 kg/10^3 g) = 0.010 kg H_2O
(0.0010 mol CH_3OH)/(0.010 kg H_2O) = **0.10 m CH_3OH**
c. (1.0 g $C_{10}H_8$)(1 mol/128.2 g) = 0.0078 mol $C_{10}H_8$
(1.0 mol C_6H_6)(78.11 g/1 mol) = 78.11 g C_6H_6
(0.0078 mol $C_{10}H_8$)/(0.07811 kg C_6H_6) = **0.10 m naphthalene**

CHAPTER NINE

9.21 The molar mass of $MgCl_2$ is 95.21 g/mol. Therefore the molarity of the solution is [0.100 g/95.21 g/mol]/0.1500 L = **7.00×10^{-3} M**

9.23 Molarity is moles of solute per liter of solution, n/V. The number of moles needed is:
n = (0.100 moles/L)(0.7500 L) = 0.0750 mol
Therefore, the required mass is
m = nM_{AgNO_3} = (0.0750 mol)(169.9 g/mol)
= **12.7 g**

9.25 The total vapor pressure is given by Raoult's Law:
$P = X_1 P_1^o + X_2 P_2^o$
Glucose is non-volatile, so its vapor pressure P_2^o is essentially zero. The mol fraction X_1 of water is n/n_{total} = 1.65/1.89 = 0.917
Then $P = X_1 P_1^o$ = (0.917)(149.38) = **137 torr**

9.27 a. (1.00 g $C_{10}H_8$)(1 mol/128.17 g) = 0.00780 mol $C_{10}H_8$
(10.0 g C_6H_6)(1 mol/78.11 g) = 0.128 mol C_6H_6
$X_{C_{10}H_8}$ = (0.00780)/(0.00780 + 0.128) = **0.0574**
Molality = moles solute/kilograms solvent
(0.00780 mol $C_{10}H_8$)/(0.0100 kg C_6H_6) = **0.780 m naphthalene**
b. From Raoult's Law, we have:
$P_{solution} = X_{solvent} P^o_{solvent}$ = (1 – 0.0574)(100. torr) = **94.3 torr**

9.29 Freezing point depression ΔT is given by
$\Delta T = K_f m$ where m is the molality (moles solute/kg solvent) of the solute.
m = (0.025 mol)/(0.0150 kg H_2O) = 1.667
ΔT = (–1.86°/m)(1.667) = **–3.1°C**

9.31 In order to calculate the freezing point depression, we must first calculate the molality of the solute in the solution.
(13.10 mg solute)(1 mmol/178.24 mg)(1 mol/10^3 mmol) = 7.350×10^{-5} mol solute
(111.4 mg camphor)(1 kg/10^6 mg) = 1.114×10^{-4} kg camphor
molality = (7.350×10^{-5} mol)/(1.114×10^{-4} kg) = 0.6598 m
Now we can make use of the equation $\Delta T_f = K_f m$, where
ΔT_f = the change in freezing point from that of pure camphor, K_f = the freezing point depression constant for camphor and m = the molality of the solute.
ΔT_f = (–40.0°C/m)(0.6598 m) = **–26.4°C**.

CHAPTER NINE

9.33 The molar mass of glucose is 180.2 g/mol. The molality of a solution is moles solute/kg solvent.
Molality = [(10. g/180.2 g/mol)/0.150 kg H_2O] = 0.37 m.
$\Delta T_f = K_f m$ = (–1.86 °C/m)(0.37 m) = –0.69°C
T_f = **–0.69°C**

9.35 In order to find the molar mass of the solute, we need the molality of the solute as well as the number of kilograms of water.
(30. mL H_2O)(1.0 g/mL)(1 kg/10^3 g) = 0.030 kg H_2O
ΔT_b = 0.52°C = (0.52°C/m)(molality); molality of solute = 1.0 m
1.0 m solute = n(solute)/(0.030 kg H_2O); n(solute) = 0.030 mol
M = m/n = (3.0 g)/(0.030 mol) = **1.0 × 10^2 g/mol**

9.37 We do not have to do unit conversions associated with the 4-gallon volume. The freezing point depression will be the same no matter what the total volume of the H_2O/ethylene glycol mixture is, as long as we take equal volumes of the two liquids. For convenience, we will work with 100.0 mL solution: 50.0 mL H_2O and 50.0 mL $(CH_2OH)_2$. We need to know the molality of ethylene glycol in order to solve the problem so we will calculate the kilograms of water and the moles of ethylene glycol.
(50.0 mL H_2O)(1.00 g/mL)(1 kg/10^3 g) = 0.0500 kg H_2O
(50.0 mL $C_2H_4(OH)_2$)(1.115 g/mL)(1 mol/62.07 g) = 0.898 mol $C_2H_4(OH)_2$
Molality of ethylene glycol = (0.898 mol)/(0.0500 kg H_2O) = 18.0 m solute
$\Delta T_f = K_f m$ = (–1.86°C/m)(18.0 m) = –33.5°C
T_f = 0.0°C – 33°C = –33.5°C
Freezing will become a problem in the car at **–33.5°C**

9.39 a. In order to find the vapor pressure of water in the solution, we need to know the mole fraction of water in the solution:
(300.g urea)(1 mol/60.06 g) = 5.00 mol urea
(1000.g H_2O)(1 mol/18.02 g) = 55.5 mol H_2O
X_{H2O} = 55.5 mol/60.5 mol = 0.917; $P°_{H2O}$ at 0°C = 4.58 torr
$P_{H2O} = X_{H2O}P°_{H2O}$ = (0.917)(4.58 torr) = **4.20 torr at 0°C**.
b. At the normal boiling point of water (100°C), the vapor pressure of water is equal to standard atmospheric pressure: $P°_{H2O}$ at 100°C = 760. torr.
$P_{solution} = X_{H2O}P°_{H2O}$ = (0.917)(760. torr) = **697 torr** at 100°C.
c. We have 1.000 kg H_2O, so the molality of urea is (5.00 mol)/(1.000 kg) = 5.00 m.
$\Delta T_b = K_b m$ = (0.52°C/m)(5.00 m) = 2.6°C, so T_b = **102.6°C**.
d. $\Delta T_f = K_f m$ = (–1.86°C/m)(5.00 m) = –9.30°C, so T_f = **–9.30°C**

CHAPTER NINE

9.41 a. First, we will use the naphthalene information to calculate the freezing point depression constant for camphor.
(0.412 g $C_{10}H_8$)(1 mol/128.17 g) = 0.00321 mol $C_{10}H_8$
(0.00321 mol $C_{10}H_8$)/(0.010 kg camphor) = 0.321 m naphthalene
$\Delta T_f = K_f m$; $K_f = \Delta T_f/m = (-13°C)/(0.321$ m$) =$ **–40.5°C/m = K_f of camphor.**

b. Now we can use the freezing point depression constant from part (a) in order to find the molar mass of the unknown substance.
$\Delta T_f = K_f m$; molality $= \Delta T_f/K_f = (-9.5°C)/(-40.5°C/m) = 0.23$ m
Molality = 0.23 = (moles solute)/(kg solvent)
n_{solute} = (0.23 m)(0.00855 kg) = 0.00201 mol
M = m/n = (1.00 g)/(0.00201 mol) = **5.0×10^2 g/mol**

9.43 Osmotic pressure π is given by π = nRT/V where n is the number of moles of solute. Then π = (0.15 mol)(0.08206 L·atm/mol·K)(298K)/0.550 L = **6.7 atm**

9.45 Glucose is $C_6H_{12}O_6$ and its molar mass is 180.2 g/mol. In order to find the osmotic pressure, we need to know the concentration (molarity) of the glucose solution.
(20. g glucose)(1 mol/180.2 g) = 0.11 mol glucose
c = (0.11 mol)/(1.0 L) = 0.11 M glucose
π = cRT = (0.11 mol/L)(0.08206 L·atm/mol·K)(300.K) = **2.7 atm**.

9.47 a. First, calculate the concentration of the protein by relationship π = cRT:
(9.25/760) atm = c(0.08206 L·atm/mol·K) (298K)
c = 4.98 × 10⁻⁴ M
This means that every liter of solution contains 4.98 × 10⁻⁴ mol protein.
M = m/n = 10. g/4.98 × 10⁻⁴ mol = **2.0×10^4 g/mol**

b. (10.0 g protein)(1 mol/2.0 × 10⁴ g) = 5.0 × 10⁻⁴ mol protein
ΔT_b = (0.52 °C/m)(5.0 × 10⁻⁴ mol protein/kg water) = **2.6×10^{-4} °C**
ΔT_f = (–1.86 °C/m)(5.0 × 10⁻⁴ mol protein/kg water) = **-9.3×10^{-4} °C**

c. Because of the difficulty in obtaining temperature measurements precise enough to even record such small ΔT's, **osmotic pressure** is the most accurate technique of the three for determining molar masses of such large particles.

9.49 When dissolved in water, the three salts dissociate as follows:
$Na_2SO_4 \rightarrow 2Na^+ + SO_4^{2-}$ i = 3
$KBr \rightarrow K^+ + Br^-$ i = 2
$CaI_2 \rightarrow Ca^{2+} + 2I^-$ i = 3

CHAPTER NINE

9.51   Molar mass of $CaCl_2$ = 110.98 g/mol
Molality = (0.100 g/110.98 g/mol)/0.5000 kg H_2O = 1.80×10^{-3} m
Since $CaCl_2$ dissociated into one Ca^{2+} and $2Cl^-$ ions, we must multiply the molality by i = 3 to obtain the freezing point depression.
$\Delta T_f = iK_fm$ = (3)(–1.86 C/m)(1.80×10^{-3} m) = -1.01×10^{-2} °C
T_f = **-1.01×10^{-2} °C**

9.53   The van't Hoff i factor is a measure of how much a particular solution deviates in its colligative properties from a solution of a non-volatile, non-dissociating solute.
i = $\Delta T_f/K_fm$ = (–3.27°C)/[(–1.86°C/m)(0.50 m)] = **3.5**

9.55   The pressure of each vapor is given by Raoult's law: p = XP°, where X is molfraction, n/n_{total}. Here each X = 1/2. Therefore,,
$p_{benzene}$ = 1/2(75 torr) = **37.5 torr**; $p_{toluene}$ = 1/2(22 torr) = **11 torr**.

9.57   Calculate the vapor pressure of each component from Raoult's law:
$P_A = X_A P°_A$ = (1/5)(80. torr) = **16 torr** = vapor pressure of A over solution
$P_B = X_B P°_B$ = (4/5)(20. torr) = **16 torr** = vapor pressure of B over solution
The composition of the vapor is determined by Dalton's law of partial pressures.
In the vapor phase, $X_A = P_A/P_{tot}$ = (16 torr)/(32 torr) = 0.50.
In the vapor phase, $X_B = P_B/P_{tot}$ = (16 torr)/(32 torr) = 0.50.
The vapor contains a **50:50 mole ratio of A:B**.

9.59   a. We know that the total pressure over the solution (66.1 torr) is equal to the sum of the vapor pressures of the two components of the solution:
66.1 torr = $X_{ethyl\ alcohol}$(44.5 torr) + (1 – $X_{ethyl\ alcohol}$)(88.7 torr)
–22.6 torr = –44.3 $X_{ethyl\ alcohol}$
So $X_{ethyl\ alcohol}$ = **0.51** and $X_{methyl\ alcohol}$ = **0.49**
b. The composition of vapor in equilibrium with this solution can be calculated from Raoult's law and Dalton's law of partial pressures:
$P_{ethyl\ alcohol}$ = (0.51)(44.5 torr) = 22.7 torr
$P_{methyl\ alcohol}$ = (0.49)(88.7 torr) = 43.5 torr
In the vapor phase, $X_{ethyl\ alcohol}$ = 22.7 torr/66.2 torr = 0.34 and
$X_{methyl\ alcohol}$ = 43.5 torr/66.2 torr = 0.66. Note that the vapor is enriched in the more volatile component (methyl alcohol).

CHAPTER NINE

9.61 Applying Raoult's law and Dalton's law to solution 1, we find:
$1.0 \text{ atm} = 0.25 P°_A + 0.75 P°_B$
Applying Raoult's law and Dalton's law to solution 2, after adding C, we find:
$1.0 \text{ atm} = 0.20 P°_A + 0.20 P°_B + 0.60(0.80 \text{ atm})$
Solve the first equation for $P°_A$: $4(1.0 \text{ atm} - 0.75 P°_B) = P°_A$
Now substitute this solution for $P°_A$ into the second equation:
$1.0 \text{ atm} = (0.20)(4)(1.0 \text{ atm} - 0.75 P°_B) + 0.20 P°_B + 0.48 \text{ atm}$
$-0.28 \text{ atm} = -0.40 P°_B$, so $\mathbf{P°_B = 0.70 \text{ atm}}$ and $\mathbf{P°_A = 1.90 \text{ atm}}$

9.63 a. From Table 9.2, the vapor pressure of H_2O at 20°C is 17.54 torr.
b. RH = 10.0 torr/17.54 torr = 0.57 or 57%
$n = PV/RT = [(10.0 \text{ torr}/760 \text{ torr/atm})(1.00 \text{ L})]/[(0.08206 \text{ L·atm/mol·K})(293.15 \text{K})]$
$= 5.47 \times 10^{-4} \text{ mol}$
$(5.47 \times 10^{-4} \text{ mol})(18.02 \text{ g/mol}) = \mathbf{9.86 \times 10^{-3} \text{ g } H_2O/L \text{ of air.}}$

9.65 In order to calculate the mole fraction of acetic acid, we need to know the number of moles of acetic acid and of water.
(50.0 g acetic acid)(1 mol/60.06 g) = 0.833 mol acetic acid
(500. g H_2O)(1 mol/18.02 g) = 27.7 mol H_2O
$n_{tot} = 0.833 \text{ mol} + 27.7 \text{ mol} = 28.5 \text{ mol}$
$X_{\text{acetic acid}} = 0.833 \text{ mol}/28.5 \text{ mol} = \mathbf{0.0292}$
$X_{H_2O} = 27.7 \text{ mol}/28.5 \text{ mol} = \mathbf{0.972}$
The molality of the acetic acid can be calculated by molality = $n_{\text{acetic acid}}/(\text{kg water})$.
Molality = 0.833 mol/0.500 kg = **1.67 m**

9.67 In order to find the mole fraction of hydrogen, we need to find the number of moles of both hydrogen and water in the solution.
(0.000164 g H_2)(1 mol/2.016 g) = 8.13×10^{-5} mol H_2
(100. g H_2O)(1 mol/18.02 g) = 5.55 mol H_2O
$n_{tot} = 5.55 \text{ mol} + 8.13 \times 10^{-5} \text{ mol} = 5.55 \text{ mol}$
$X_{H_2} = 8.13 \times 10^{-5} \text{ mol}/5.55 \text{ mol} = 1.46 \times 10^{-5}$
According to Henry's law, $p_{H_2} = K_H X_{H_2}$, where p_{H_2} is the partial pressure of hydrogen above the solution. In this problem, $p_{H_2} = 1.00$ atm.
$K_H = (1.00 \text{ atm})/(1.46 \times 10^{-5}) = \mathbf{6.85 \times 10^4 \text{ atm}}$

9.69 First we will calculate the concentration of the solution by making use of the relationship $\pi = cRT$.
$(91.2/760) \text{ atm} = c(0.08206 \text{ L·atm/mol·K})(298\text{K})$

c = 4.91 × 10⁻³ M

Assuming that the volume of the water did not change when the protein was added, we can use this concentration to calculate the number of moles of protein in the solution.

4.91×10^{-3} mol/L = $n_{protein}$/0.010 L; $n_{protein}$ = 4.91×10^{-5} mol

M = m/n = 0.65 g/4.91×10^{-5} mol = **1.3×10^4 g/mol**

9.71 a. For an aqueous solution, $\Delta T_f = iK_f m$, so
$i = \Delta T_f/K_f m = (-0.0744°C)/[(-1.86°C/m)(0.010\ m)] = 4.0$

b. $i = \Delta T_b/K_b m$ so $\Delta T_b = iK_b m = (4.0)(0.52°C/m)(0.010\ m) = 0.021°C$

Since we are working with an aqueous solution, this means that the actual boiling point of the solution is 100.021°C.

c. A van't Hoff i factor of 4 means that the $TiCl_3$ dissociates completely into four fragments (ions) when dissolved in water. These must be Ti^{3+} and $3Cl^-$

9.73 Applying Raoult's law and Dalton's law to the initial solution, we find:
250. torr = 0.33 $P°_A$ + 0.67$P°_B$

Applying Raoult's law and Dalton's law to the solution after more A has been added, we find:

  300. torr = 0.50$P°_A$ + 0.50$P°_B$

Solving the second equation for $P°_B$, we get:

  2(300. torr − 0.50 $P°_A$) = $P°_B$

Then we plug the solution for $P°_B$ into the first equation to get:

  250. torr = 0.33$P°_A$ + (0.67)(2)(300. torr − 0.50$P°_A$)

  −150 torr = −0.33$P°_A$, so at 50°C

$P°_A$ = 450 torr and $P°_B$ = 150 torr.

9.75 a. This is a graph of vapor pressure for each of the components vs. X_{CHCl_3}:

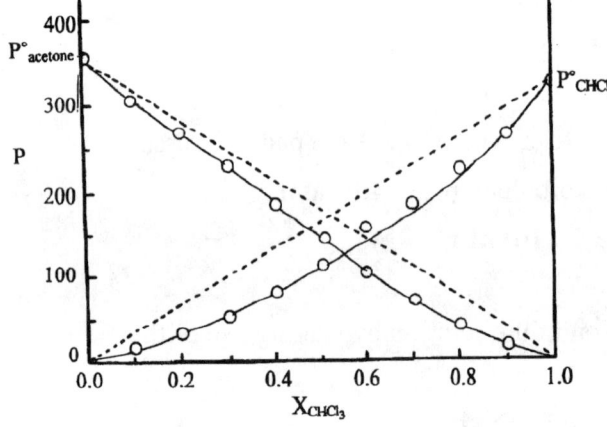

CHAPTER NINE

b. Vapor pressure of each component vs. X_{CHCl_3}. c. The top line is the total pressure.

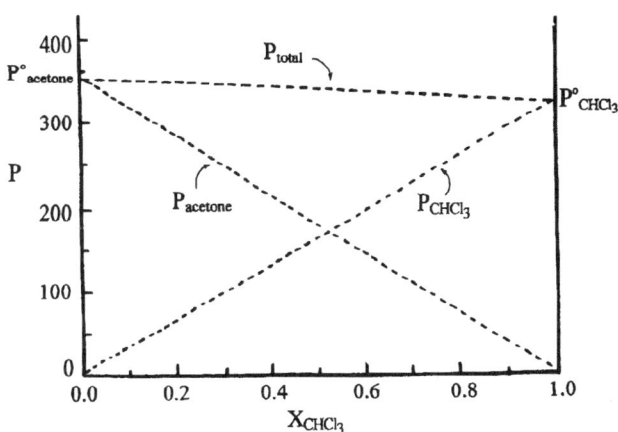

d. All of the dashed lines are linear, and correspond to ideal solution behavior according to Raoult's law. The actual vapor pressures exhibit negative deviations from Raoult's law: the data points are uniformly below the ideal lines. As each component's molfraction approaches 1, its vapor pressure approaches that predicted by Raoult's law.

9.77   At 20°C, the vapor pressure of water is 17.54 torr (Table 9.2). Thus, the partial pressure of the displaced air, collected over water, is 732.5 torr (or 0.964 atm). We can use this information to calculate the moles of air displaced (which is equal to the moles of unknown vapor produced):
n = PV/RT = (0.964 atm)(0.0400 L)/(0.08206 L·atm/mol·K)(293K) = 1.60×10^{-3} mol
Now we can calculate the molar mass of the unknown liquid:
M = m/n = (0.300 g)/(1.60×10^{-3} mol) = **188 g/mol**

9.79   A 2.00- L ethanol solution contains 1.00 L H_2O and 1.00 L C_2H_5OH.
   (1.00 L C_2H_5OH)(790 g/L) = 790 g ethanol
   (2.00 L solution)(940 g/L) = 1880 g solution
   1880 g solution − 790 g ethanol = 1090 g H_2O
   (790 g ethanol)(1 mol/46.07 g) = 17 mol ethanol
   (1090 g H_2O)(1 mol/18.02 g) = 60.5 mol H_2O
Molality of ethanol = (17 mol ethanol)/(1.09 kg H_2O) = **16 m ethanol**
$X_{ethanol}$ = (17 mol)/(77.5 mol) = **0.22**
X_{H_2O} = (60.5 mol)/(77.5 mol) = **0.78**
Molarity of ethanol = 17 mol/2.00 L = **8.5 M**

CHAPTER NINE

$\Delta T_f = K_f m = (-1.86°C/m)(16\ m) = -30.°C$, $\mathbf{T_f = -30.°C}$, assuming that ethanol is a non-volatile solute, which is probably not a bad assumption at that temperature. You do not have to worry about this solution freezing at $-10°C$.

9.81 According to Henry's law, $p = K_H X$, where p = partial pressure of the gas above the liquid solvent, K_H = the Henry's law constant and X = the mole fraction of the gas dissolved in the liquid. In order to calculate the Henry's law constant, then, we need to know the mole fraction of helium dissolved in water.
$n = pV/RT = (0.96\ atm)(0.0093\ L)/(0.08206\ L·atm/mol·K)(293K)$
$= 3.7 \times 10^{-4}$ mol He
$(100.\ g\ H_2O)(1\ mol/18.02\ g) = 5.55$ mol H_2O
$X_{He} = 3.7 \times 10^{-4}\ mol/5.55\ mol = 6.7 \times 10^{-5}$
$K_H = p/X = 0.96\ atm/6.7 \times 10^{-5} = \mathbf{1.4 \times 10^4\ atm}$

9.83 a. The critical temperature is approximately 450K.
b. The normal boiling point is approximately 375K.
c. The normal melting point is approximately 340K.
d. At a normal room temperature of *ca*. 300K, the equilibrium vapor pressure of the solid is approximately 0.4 atm.

9.85 We will work with 1000. g of solution which contains 980. g H_2SO_4 and 20. g H_2O. In order to calculate the molality of the solution, we need to find the number of moles of H_2SO_4 and the number of kilograms of H_2O.
$(980.\ g\ H_2SO_4)(1\ mol/98.07\ g) = 9.99$ mol H_2SO_4
$(20.\ g\ H_2O)(1\ kg/10^3\ g) = 0.020$ kg H_2O
Molality = $(9.99\ mol\ H_2SO_4)/(0.020\ kg\ H_2O) = \mathbf{500\ m\ H_2SO_4}$

9.87 NaCl costs half as much for any given mass, say 100. g, than $CaCl_2$. Now we need to know how many moles of ions 100. g of each of the two salts will produce.
$(100.\ g\ NaCl)(1\ mol\ NaCl/58.44\ g)(2\ mol\ ions/mol\ NaCl) = 3.42$ mol ions
$(100.\ g\ CaCl_2)(1\ mol\ CaCl_2/111.0\ g)(3\ mol\ ions/mol\ CaCl_2) = 2.70$ mol ions
So NaCl produces $(3.42/2.70) = 1.27$ times the ions at half the cost. It is **2.54 times** as cost-effective as $CaCl_2$ at melting ice and snow in winter.

9.89 We are given that the Henry's law constant for CO_2 is 1.26×10^6 torr. At a pressure of 1.5 atm, the mole fraction of CO_2 in the solution is $(1.5 \times 760$ torr$)/(1.26 \times 10^6$ torr$) = 9.05 \times 10^{-4}$. At a pressure of 1.0 atm, the mole fraction of CO_2 in the solution is $(1.0 \times 760$ torr$)/(1.26 \times 10^6$ torr$) = 6.03 \times 10^{-4}$. If we

CHAPTER NINE

assume that the density of the soda is 1.0 g/cm^3 and that essentially all of the mass of the soda is due to water, then we know that there are 1000 g of water in the solution both before and after the escape of CO_2.

(1000 g H_2O)(1 mol/18.02 g) = 55.5 mol H_2O

n_{CO_2} at 1.5 atm = (9.05 × 10^{-4})(55.5 mol) = 0.050 mol CO_2

n_{CO_2} at 1.0 atm = (6.03 × 10^{-4})(55.5 mol) = 0.033 mol CO_2

This means that approximately 0.017 mol CO_2 escaped from the soda when the pressure was released. Under the given conditions, this quantity of CO_2 will occupy (0.017 mol)(0.08206 L·atm/mol·K)(293K)/(1.0 atm) = **0.41 L**

9.91 Seawater is a 0.52 M salt solution, which we are taking to be 0.52 M NaCl. For this salt, i = 2, so that π = icRT = 2(0.52 mol/L)(0.08206 L·atm/mol·K)(298K) = 25 atm. So a pressure greater than 25 atm will be required in order to bring about reverse osmosis.

Since seawater is an aqueous solution, we will assume that molality is numerically identical to molarity, 0.52 m. Now we can calculate the freezing point of seawater:

ΔT_f = iK_fm = 2(−1.86°C/m)(0.52 m) = −1.9°C, so T_f = **−1.9°C**

9.97 a. As seen in the figure, at room temperature and pressure, graphite, and not diamond, is the stable form of carbon. Diamonds are *not* "forever" at room temperature and pressure. Fortunately, the process of diamond changing to graphite is very slow.

b. As seen in the figure, the transition of diamond to a liquid state at 1250K occurs at about 900,000 atm.

c. Graphite will melt at 2500K. As the pressure increases, the liquid will solidify to graphite. At point A the pressure causes the graphite to change into diamond. At point B, the diamond will melt into liquid carbon.

d. Point C is a triple point, where graphite, diamond and liquid carbon coexist.

9.99 The equation for osmotic pressure, πV=nRT, can be related to the mass m and molar mass M of the solute by πV= (m/M)RT or π = (m/V)RT/M.

Thus, we see that the osmotic pressure is directly proportional to the solute concentration m/V, expressed in grams per liter. If you were to graph π vs. m/V, the result would be a straight line with a slope equal to RT/M. Thus, knowing the slope at a particular temperature would give the molar mass of the solute.

CHAPTER TEN

Chemical Equilibrium

10.1 Reactants ↔ Products and heat
If the temperature is raised, the reaction will shift to the left so that the system can absorb heat and reestablish equilibrium.

10.3 Increased pressure will shift the equilibrium to the direction with the smaller number of moles of gases.
    a. $K_p = p^2_{NO_2}/(p_{O_2}p^2_{NO})$
       A pressure increase will shift the equilibrium to the **right**.
    b. $K_p = p_{N_2O}\, p_{NO_2}/p^3_{NO}$
       A pressure increase will shift the equilibrium to the **right**.

10.5 Increased pressure will shift the equilibrium to the direction with the smaller number of moles of gases.
    a. $K_p = p_{CO}$
       An increase in the partial pressure of CO will shift the equilibrium to the **left**.
    b. $K_p = p_{Cl_2}\, p_{PCl_3}/p_{PCl_5}$
       An increase in the partial pressure of PCl_3 will shift the equilibrium to the **left**.

10.7 Pure solids or liquids do not appear in the equilibrium expressions.
    a. $K = [Zn^{2+}]p_{H_2}/[H^+]^2$
    b. $K = [F^-]^2[Ca^{2+}]$
    c. $K = 1/[Cl^-]^2[Cu^{2+}]$
    d. $K = [V^{2+}]^3/[Cr^{3+}]^2$

10.9 a. Addition of chlorine will shift the equilibrium to the **left.**
     b. Addition of oxygen will shift the equilibrium to the **right**
     c. A decrease in temperature will shift the equilibrium to the **left.**
     d. An increase in pressure will shift the equilibrium to the **right**.
     e. A decrease in volume will shift the equilibrium to the **right.**
     f. Addition of a catalyst will leave the equilibrium position **unchanged.**

10.11 a. $K_2 = (K_1)^{1/2}$
      b. $K_2 = 1/K_1$

10.13 $K = K_2/K_1 = (1.66 \times 10^{12})/(1.49 \times 10^{13}) = \mathbf{0.111}$

CHAPTER TEN

10.15 We know that $K_p = K_c(RT)^{\Delta n}$ for gas-phase reactions. For this particular reaction, $\Delta n = 2 - 1 = 1$.
∴ $2.4 = K_c(0.08206)(373)$.
$K_c = 0.078$

10.17 We are given that $K = 2.4 \times 10^{-5}$ and $Q = 2.4 \times 10^{-6}$
$Q < K$, so the system is **not at equilibrium** and the reaction will proceed spontaneously **to the right.**

10.19 We must find the value for the reaction quotient, Q_p, and determine whether it is greater than, less than or equal to the equilibrium constant, K_p.
$Q_p = p_{COCl_2}/(p_{CO}p_{Cl_2}) = (0.12)/(0.52)(0.35) = 0.66 > K_p$
Since $Q > K$, the reaction is **not at equilibrium** and the reaction will **proceed spontaneously to the left.**

10.21 $K_3 = K_1(K_2)^2 = (3.8)(0.70)^2 = \mathbf{1.9}$
We must find Q_p and compare it to K_p.
$Q_p = p^2_{CO}/p_{CO_2} = (1.50)^2/(1.40) = 1.6 < K_p$
Since $Q < K$, the system is not at equilibrium, and the reaction will proceed **to the right**.

10.23 For the reaction given, $K_p = 0.01 = p_{CO_2} p_{H_2}/(p_{CO} p_{H_2O})$.
$0.01 = (0.1 \text{ atm}) p_{H_2}/[0.1 \text{ atm}](0.1 \text{ atm})$
$p_{H_2} = \mathbf{10\ atm}$

10.25 Let the reaction occur to an extent "x" in order to reach equilibrium. Since we are starting with 0.012500 atm PCl_5, this means that $p_{Cl_2} = x$, $p_{PCl_3} = x$ and $p_{PCl_5} = (0.012500 - x)$ once equilibrium is achieved. Now we can set up the equilibrium expression and solve for "x":
$K = 2.15 = x^2/(0.012500 - x)$, where x = pressure of PCl_3 or Cl_2 formed.
$x^2 = 0.026875 - 2.15x$; $x^2 + 2.15x - 0.026875 = 0$
$x = \frac{1}{2}[-2.15 \pm \sqrt{2.15^2 + 4(1)(0.026875)}]$
x = 0.012428 atm (plus a negative root which will not yield a physically reasonable answer)
$p_{PCl_3} = p_{Cl_2} = \mathbf{0.012428\ atm}$
$p_{PCl_5} = 0.012500 - 0.012428 = \mathbf{7.2 \times 10^{-5}\ atm}$

CHAPTER TEN

10.27 Let the reaction occur to an extent "x" in order to reach equilibrium. Since we are starting with 0.100 atm SO_2 and 0.100 atm O_2, this means that $p_{SO_3} = 2x$, $p_{O_2} = 0.100 - x$ and $p_{SO_2} = (0.100 - 2x)$ once equilibrium is achieved. Now we can set up the equilibrium expression and solve for "x".

$K = 0.587 = (2x)^2/(0.100 - x)(0.100 - 2x)^2$, where $2x$ = pressure of SO_3 formed.

Since we are going to end up with an "x^3"-term in the denominator of the equilibrium expression, we should look for some assumptions that will simplify the math. Let us assume that "x" is very small in comparison with 0.100. Then we can write $(0.100 - x) \approx 0.100$ and $(0.100 - 2x) \approx 0.100$.

$0.587 \approx 4x^2/(0.100)^3$; $4x^2 \approx 5.87 \times 10^{-4}$; $x \approx 0.0121$

Our assumption was not very good, but we can remedy the situation by continuing to solve the equation by iteration. This means that we will take our initial answer for "x", simplify the denominator with this new value and solve for "x" again. We will continue to do this until we get the same value for "x" twice in a row. Here we go...

$(0.100 - x) \approx 0.0879$, $(0.100 - 2x) \approx 0.0758$
$(0.587) \approx 4x^2/(0.0879)(0.0758)^2$; $x \approx 8.61 \times 10^{-3}$

$(0.100 - x) \approx 0.091$, $(0.100 - 2x) \approx 0.083$
$(0.587) \approx 4x^2/(0.091)(0.083)^2$; $x \approx 4.99 \times 10^{-3}$

$(0.100 - x) \approx 0.095$, $(0.100 - 2x) \approx 0.090$
$(0.587) \approx 4x^2/(0.095)(0.090)^2$; $x \approx 0.011$

$(0.100 - x) \approx 0.089$, $(0.100 - 2x) \approx 0.078$
$(0.587) \approx 4x^2/(0.089)(0.078)^2$; $x \approx 8.9 \times 10^{-3}$

$(0.100 - x) \approx 0.091$, $(0.100 - 2x) \approx 0.082$
$(0.587) \approx 4x^2/(0.091)(0.082)^2$; $x \approx 9.5 \times 10^{-3}$

$(0.100 - x) \approx 0.090$, $(0.100 - 2x) \approx 0.081$
$(0.587) \approx 4x^2/(0.090)(0.081)^2$; $x \approx 9.3 \times 10^{-3}$

$(0.100 - x) \approx 0.091$, $(0.100 - 2x) \approx 0.081$
$(0.587) \approx 4x^2/(0.091)(0.081)^2$; $x \approx 9.4 \times 10^{-3}$

$(0.100 - x) \approx 0.091$, $(0.100 - 2x) \approx 0.081$
$(0.587) \approx 4x^2/(0.091)(0.081)^2$; $x \approx 9.4 \times 10^{-3}$

$x = 9.4 \times 10^{-3}$ atm, **$p_{SO_3} = 0.019$ atm, $p_{O_2} = 0.091$ atm and $p_{SO_2} = 0.081$ atm.**

CHAPTER TEN

10.29 First, calculate Q_p and compare it with K_p in order to determine whether the reaction will proceed to the right or to the left.

$Q_p = p_{H_2}p_{CO_2}/p_{H_2O}p_{CO} = (0.100)^2/(0.100)^2 = 1$

$Q_p < K_p$ so the reaction will proceed to the right.

Let the reaction occur to an extent "x" in order to reach equilibrium. Since we are starting with 0.100 atm each of CO, H_2O, CO_2 and H_2 and the reaction is proceeding toward the right, this means that $p_{CO} = p_{H_2O} = (0.100 - x)$, $p_{CO_2} = p_{H_2} = (0.100 + x)$ once equilibrium is achieved. Now we can set up the equilibrium expression and solve for "x".

$3.9 = (0.100 + x)^2/(0.100 - x)^2$

Expanding the squared terms would give a quadratic equation, which could be solved in the usual manner. But an easier solution is possible by recognizing that the equation is a perfect square. Taking the square root of both sides yields $\sqrt{3.9} = 1.97 = (0.100 + x)/(0.100 - x)$ which is easily solved: $x = 0.033$.

Thus, **$p_{CO_2} = p_{H_2} = 0.133$ atm, and $p_{CO} = p_{H_2O} = 0.067$ atm**

10.31 a. $K = p_{H_2}p_{CO_2}/p_{H_2O}p_{CO}$

b. For this reaction, $\Delta n = 0$ so $K_c = K_p = 0.10$.

If we let x = number of moles of CO and H_2 formed in order to achieve equilibrium, then $[CO_2] = [H_2] = x/V$ and $[CO] = [H_2O] = (1 - x)/V$ at equilibrium.

$K_c = 0.10 = (x/V)^2/((1 - x)/V)^2 = x^2/(1 - x)^2$

We recognize this as a perfect square, so we can take the square root of both sides:
$\sqrt{0.10} = 0.316 = x/(1 - x)$

Now solve for x: $x = 0.24$

At equilibrium, there are **0.24 moles each of CO_2 and H_2, and there are 0.76 moles each of CO and H_2O.**

c. If equimolar concentrations of all reactants and products are present in this reaction, $Q = 1$. $Q > K$, and the reaction will proceed to the left. **The reaction will favor CO formation.**

10.33 a. Calculate the initial concentration of CO_2: $[CO_2]_o = 7.50$ mol/5.00 L = 1.50 M. We know that there was 20.0% dissociation which means that 0.300 M CO_2 was consumed, leaving 1.20 M CO_2, 0.300 M CO and 0.150 M O_2 at equilibrium.

$K_c = [O_2][CO]^2/[CO_2]^2 = (0.150)(0.300)^2/(1.20)^2 = \mathbf{9.38 \times 10^{-3}}$

$K_p = K_c (RT)^{\Delta n} = (9.38 \times 10^{-3})(RT)^1 = (9.38 \times 10^{-3})(0.08206)(423) = \mathbf{0.326}$

For the reverse reaction, the equilibrium constants will be the reciprocals of those for the forward reaction: $K_c' = 1/K_c = \mathbf{107}$ and $K_p' = 1/K_p = \mathbf{3.07}$

b. An increase in pressure will shift the equilibrium to the **left**.

63

CHAPTER TEN

c. A decrease in temperature will shift the equilibrium to the **left**, which is the direction that absorbs heat.

10.35 $AgI(s) \rightarrow Ag^+(aq) + I^-(aq)$
Let x moles of AgI dissolve to form 1 L of saturated solution.
$K_{sp} = 8.5 \times 10^{-17} = [Ag^+][I^-] = x^2$
$x = 9.2 \times 10^{-9}$ moles
9.2×10^{-9} moles AgI dissolved to form 1 L saturated solution
Solubility = 9.2×10^{-9} moles/L

10.37 (0.062 g CuCl)(1 mol/98.999 g) = 6.3×10^{-4} moles CuCl dissolved in 1.0 L
6.3×10^{-4} moles of Cu^+ and 6.3×10^{-4} moles of Cl^- were produced per liter.
$K_{sp}(CuCl) = [Cu^+][Cl^-] = (6.3 \times 10^{-4})^2 =$ **4.0×10^{-7}**

10.39 $NiS(s) \rightarrow Ni^{2+}(aq) + S^{2-}(aq)$
K_{sp} for NiS = 1.4×10^{-24}. Dissolving 0.010 M Na_2S yields 0.010 M S^{2-}. If we let the NiS dissolve to an extent "x" in the Na_2S solution, then at equilibrium $[Ni^{2+}]$ = x, and $[S^{2-}]$ = x + 0.010.
$K_{sp} = 1.4 \times 10^{-24} = (x)(x + 0.010) \approx (x)(0.010)$
$x \approx 1.4 \times 10^{-22}$
Check: $(1.4 \times 10^{-22} + 0.010) = 0.010$, so the approximation is valid.
x = **molar solubility of NiS = 1.4×10^{-22} mol/L**
$[Ni^{2+}] = 1.4 \times 10^{-22}$ M; $[S^{2-}] = 0.010$ M.

10.41 From Table 10.4, K_{sp} for AgBr = $5.2 \times 10^{-13} = [Ag^+][Br^-]$.
$[Ag^+] = K_{sp}/[Br^-] = 5.2 \times 10^{-13}/1.2 \times 10^{-8}$
$[Ag^+] = 4.3 \times 10^{-5}$ M

10.43 K_{sp} for $PbCO_3 = 1.5 \times 10^{-13}$, and K_{sp} for $MgCO_3 = 4.0 \times 10^{-5}$.
At the point where $MgCO_3$ begins to precipitate, $[Mg^{2+}][CO_3^{2-}] = 4.0 \times 10^{-5}$.
(0.10 M) $[CO_3^{2-}] = 4.0 \times 10^{-5}$ so **$[CO_3^{2-}] = 4.0 \times 10^{-4}$ M.**
At this point, $[Pb^{2+}][CO_3^{2-}] = 1.5 \times 10^{-13}$ so $[Pb^{2+}] = (1.5 \times 10^{-13})/(4.0 \times 10^{-4})$.
$[Pb^{2+}] = 3.8 \times 10^{-10}$ M

CHAPTER TEN

10.45 K_{sp} for $Mn(OH)_2) = 1.0 \times 10^{-14}$; K_{sp} for $Mg(OH)_2 = 1.2 \times 10^{-11}$.

When $Mg(OH)_2$ first precipitates, the Mn^{2+} will already have precipitated to a large extent since it is the less soluble substance. We can calculate the concentration of Mn^{2+} remaining in solution at this point from the known OH^- concentration at which Mg^{2+} begins to precipitate:

$[OH^-]^2 = K_{sp}(Mg(OH)_2)/[Mg^{2+}]$
$[OH^-] = [1.2 \times 10^{-11}/(0.10)]^{1/2} = 1.1 \times 10^{-5}$ M
$[Mn^{2+}] = K_{sp}(Mn(OH)_2)/[OH^-]^2$
$[Mn^{2+}] = 1.0 \times 10^{-14}/(1.1 \times 10^{-5})^2 = 8.3 \times 10^{-5}$ M

The fraction of Mn^{2+} remaining in solution when Mg^{2+} begins to precipitate is $(8.3 \times 10^{-5})/0.10$ M $= 8.3 \times 10^{-4}$. Therefore the fraction that has precipitated is $1 - 8.3 \times 10^{-4} =$ **0.99917**.

10.47 a. $Mg^{2+}(aq) + 2OH^-(aq) \rightarrow Mg(OH)_2(s)$
b. $3Ca^{2+}(aq) + 2PO_4^{3-}(aq) \rightarrow Ca_3(PO_4)_2(s)$

10.49 a. $K_p = p_{UF_6}/p_{F_2}$
b. $K_p = p^2_{BrCl}/p_{Cl_2}$
c. $K_p = p_{SiF_4}$

10.51 Since we are starting with pure NH_4OCONH_2, we know that NH_3 and CO_2 will form in a 2:1 mole ratio which is the same as a 2:1 ratio of partial pressures.
Let $p_{CO_2} = x$, then $p_{NH_3} = 2x$
$(2x)^2(x) = 4x^3 = K_p = 5.78 \times 10^{-5}$
$x = 2.44 \times 10^{-2}$ atm $= p_{CO_2}$
$p_{NH_3} = 4.87 \times 10^{-2}$ atm

10.53 We know that the total pressure of the system will be equal to the sum of the partial pressures of PCl_5, PCl_3 and Cl_2. If we let "x" represent the partial pressure of Cl_2 (and also that of PCl_3 because of the 1:1 mole relationship), then we can write:
$P_{tot} = p_{PCl_5} + p_{PCl_3} + p_{Cl_2}$
$0.54 = (0.30 - x) + x + x = 0.30 + x$
so $x = 0.24$ atm $= p_{PCl_3} = p_{Cl_2}$; $p_{PCl_5} = 0.30 - x = 0.06$ atm
$Q_p = (0.24)^2/(0.06) = 0.96 < K_p$
Since Q < K, the reaction is **not at equilibrium** and will proceed to the right.

CHAPTER TEN

10.55 We know that the total pressure is the sum of the partial pressures:
$P_{tot} = 0.135$ atm $= p_{N_2O_4} + p_{NO_2}$
If we let $x = p_{NO_2}$, then $(0.135 - x) = p_{N_2O_4}$.
$K_p = 8.8 = (0.135 - x)/x^2$, $8.8x^2 = 0.135 - x$, $8.8x^2 + x - 0.135 = 0$

$$x = \frac{1}{2(8.8)}\left[-1 \pm \sqrt{1^2 + 4(8.8)(0.135)}\right] = 0.0794 \text{ atm (plus a negative root which}$$

will not yield a physically reasonable answer)
$p_{NO_2} = x = \mathbf{0.079}$ **atm**, $p_{N_2O_4} = 0.135 - x = \mathbf{0.056}$ **atm**

10.57 If we let P_o = the initial pressure of $COCl_2$, and if we let x = pressure of $COCl_2$ that dissociates, then at equilibrium:
$p_{COCl_2} = P_o - x$, $p_{CO} = x$, $p_{Cl_2} = x$
$P_{tot} (P_o - x) + x + x = P_o + x = 2.0$ atm
$K_p = 8.0 \times 10^{-19} = x^2/(P_o - x)$, but $P_o = (2.0-x)$ so $(P_o - x) = (2.0 - 2x)$
$8.0 \times 10^{-19} = x^2/(2.0 - 2x)$, $x^2 = 1.6 \times 10^{-18} - 1.6 \times 10^{-18}(x)$
$x^2 + 1.6 \times 10^{-18}(x) - 1.6 \times 10^{-18} = 0$

$$x = \frac{1}{2}\left[-1.6 \times 10^{-18} \pm \sqrt{(1.6 \times 10^{-18})^2 + 4(1)(1.6 \times 10^{-18})}\right]$$

$x = 1.26 \times 10^{-9}$ (plus a negative root which will not yield a physically reasonable result)
Percent dissociation = amount dissociated/initial amount percent dissociation of
$COCl_2 = x/P_o = 1.26 \times 10^{-9}/2.00 = 6.3 \times 10^{-10} = \mathbf{6.3 \times 10^{-8}}$ **%**

10.59 $HgS(s) \rightarrow Hg^{2+}(aq) + S^{2-}(aq)$
(1 molecule)(1 mol/6.022×10^{23} molecules) = 1.66×10^{-24} mol HgS
We want to know the minimum volume required in order to dissolve 1.66×10^{-24} mol HgS. In order to calculate this volume, we need to know the concentration of a saturated solution of HgS in water.
$K_{sp} = 3 \times 10^{-53} = [Hg^{2+}][S^{2-}] = 3 \times 10^{-53} = x^2$; $x = 5.5 \times 10^{-27}$ M
5.5×10^{-27} mol/L = $(1.66 \times 10^{-24}$ mol$)$V
V = 300 L
This is a large volume!!! It is a consequence of the extremely small K_{sp}.

10.61 a. $La(IO_3)_3(s) \rightarrow La^{3+}(aq) + 3IO_3^-(aq)$
$K_{sp} = [La^{3+}][IO_3^-]^3$
If we let "x" moles of $La(IO_3)_3$ dissolve to form 1 L of saturated solution, then $[La^{3+}] = x$ and $[IO_3^-] = 3x$.
$K_{sp} = (x)(3x)^3 = 27x^4 = 6.10 \times 10^{-12}$, $x = 6.89 \times 10^{-4}$

CHAPTER TEN

So 6.89×10^{-4} mol La(IO$_3$)$_3$ dissolved to form 1 L of saturated solution.
(6.89×10^{-4} mol La(IO$_3$)$_3$ (663.61 g/mol) = 0.457 g
Solubility = 0.457 g/L
[IO$_3^-$] = (3x) mol/L = **2.07×10^{-3} M**

b. Adding lanthanum nitrate would decrease the solubility of lanthanum iodate in water due to the common ion effect.

10.63 a. The answers to this problem are independent of the volume taken, so we will perform all calculations for 1.00 L.
The mass of the gases will not change as the sample reaches equilibrium. Therefore, we can determine the initial number of moles of PCl$_5$ from the density:
(1.0 L)(3.03 g/L) = 3.03 g PCl$_5$ initially.
n° = 3.03 g/208.2 g/mol = 1.46×10^{-2} mol PCl$_5$. The initial PCl$_5$ pressure is
p° = nRT/V = (1.46×10^{-2} mol)(0.08206 L·atm/mol·K)(503K)/1.0 L
p° = 0.60 atm
Now set up the equilibrium expression based on the balanced equation. If x is the pressure change that results in equilibrium:

PCl$_5$ → PCl$_3$ + Cl$_2$

P$_{total}$ = p$_{PCl_5}$ + p$_{PCl_3}$ + p$_{Cl_2}$

1.0 atm = 0.60 − x + x + x

1.0 = 0.60 + x

x = 0.40 atm

Therefore, the fraction of the original PCl$_5$ that has dissociated is (0.40/0.60) = **2/3**.

b. If the starting pressure of PCl$_5$ was 0.60 atm and 0.40 atm dissociated, the equilibrium partial pressures of each gas are
PCl$_5$: 0.60 atm − 0.40 atm = **0.20 atm**
PCl$_3$: **0.40 atm**
Cl$_2$: **0.40 atm**

c. K$_p$ = p$_{PCl_3}$ p$_{Cl_2}$/p$_{PCl_5}$ = (0.40)(0.40)/0.20
K$_p$ = 0.80

10.65. We know that the total pressure of the system is equal to the sum of the partial pressures: 1.00 atm = p$_{NO_2}$ + p$_{N_2O_4}$. If we let x = p$_{NO_2}$, then (1.00 − x) = p$_{N_2O_4}$.
At equilibrium, K$_p$ = 8.8 = (1.00 − x)/x^2; 8.8x^2 + x − 1.00 = 0;

$$x = \frac{1}{17.6}\left[-1 \pm \sqrt{1^2 + 4(8.8)(1.00)}\right]; \quad x = 0.285 \text{ atm (plus a negative root which will}$$

not yield a physically meaningful result). So we know that before the volume change p$_{NO_2}$ = 0.285 atm and p$_{N_2O_4}$ = 0.715

CHAPTER TEN

If the volume of the container is increased to 3.00 times its original value, then instantaneously $P_{oNO_2} = 0.285$ atm/3 = 0.0950 atm and $P_{oN_2O_4} = 0.715$ atm/3 = 0.238 atm. This is not an equilibrium situation, though, and the reaction will shift to the left in order to re-establish equilibrium. (You can prove this to yourself by calculating Q at this point and showing that Q > K.)
If we let the reaction proceed to the left to an extent "x" in order to achieve equilibrium, then we can write: $8.8 = p_{N_2O_4}/p^2_{NO_2} = (0.238 - x)/(0.0950 + 2x)^2$
$8.8(9.025 \times 10^{-3} + 0.38x + 4x^2) = 0.238 - x$
$35.2x^2 + 4.344x - 0.15858 = 0$

$$x = \frac{1}{2(35.2)}\left[-4.344 \pm \sqrt{4.344^2 + 4(35.2)(0.15858)}\right] = 0.0295 \text{ atm (plus a negative root}$$

which will not yield a physically reasonable answer). So, once equilibrium is re-established, $p_{NO_2} = 0.0950 + 0.059 = \textbf{0.154 atm}$ and $p_{N_2O_4} = 0.238 - 0.030 = \textbf{0.208 atm}$

10.67 $I_2(g) \rightarrow 2I(g)$

We know that $p_{I_2} + p_I = 9.2$ atm. If we let P_o = the initial pressure of I_2 and if we let x = the partial pressure of I_2 molecules dissociated, then we can also write $(P_o - x) + 2x = 9.2$ atm and $x/P_o = 0.799$.

$P_o + x = 9.2$; $P_o = (9.2 - x)$; $x/(9.2 - x) = 0.799$
$x = 7.35 - 0.799x$; $1.799x = 7.35$; $x = 4.1$ atm
$P_o = 5.1$ atm; $p_{I_2} = 1.0$ atm; $p_I = 8.2$ atm
$K_p = p_I^2/p_{I_2} = (8.2)^2/(1.0) = \textbf{67}$

10.69 If we let P_o = the initial partial pressure of CO, then $4P_o$ = the initial partial pressure of H_2O. Initially, $P_{tot} = 1.00$ atm = $P_o + 4P_o = 5P_o$. $P_o = 0.200$ atm. If we let the reaction proceed to the right to an extent "x" in order to achieve equilibrium, then we have:

$p_{CO} = (0.200 - x)$, $p_{H_2O} = (0.800 - x)$, $p_{CO_2} = x$ and $p_{H_2} = x$
$0.160 = x^2/[(0.200 - x)(0.800 - x)]$
$0.160(0.160 - x + x^2) = x^2$
$0.0256 - 0.160x + 0.160x^2 = x^2$
$0.840x^2 + 0.160x - 0.0256 = 0$

$$x = \frac{1}{1.68}\left[-0.160 \pm \sqrt{0.160^2 + 4(0.840)(0.0256)}\right]; \quad x = 0.104 \text{ (plus a negative}$$

root which will not yield a physically meaningful result)
At equilibrium, $p_{CO} = 0.096$ atm, $p_{H_2O} = 0.696$ atm, $p_{CO_2} = 0.104$ atm and $p_{H_2} = 0.104$ atm. So the fraction of carbon monoxide converted to carbon dioxide is $0.104/0.200 = \textbf{0.520}$.

CHAPTER TEN

10.71 From Table 10.4, K_{sp} for $CaCO_3$ is $8.7 \times 10^{-9} = [Ca^{2+}][CO_3^{2-}]$
Therefore $[Ca^{2+}] = [CO_3^{2-}] = \sqrt{K_{sp}} = 9.3 \times 10^{-5}$ M
$(9.3 \times 10^{-5}$ M$)(1.0000 \times 10^4$ L$)(100.08$ g/mol$) = 93$ g
93 g of $CaCO_3$ could be dissolved in this volume of water.

10.73 Seawater contains 0.13% Mg by mass. We will convert this to g/L then moles/L.
$(0.0013)(1.025$ g/mL$)(10^3$ mL/L$) = 1.33$ g/L
1.33 g/L$/24.3$ g/mol $= 5.5 \times 10^{-2}$ M Mg
From Table 10.4, K_{sp} for $Mg(OH)_2 = 1.2 \times 10^{-11}$.
$Ca(OH)_2$ is soluble in water to the extent 0.021 mol/L, so the OH^- concentration will be $(0.021$ M$)(2$ $OH^-/Ca(OH)_2) = 0.042$ M. Then
$K_{sp} = 1.2 \times 10^{-11} = [Mg^{2+}][OH^-]^2$
$[Mg^{2+}] = 1.2 \times 10^{-11}/(0.042)^2 = 6.8 \times 10^{-9}$ M
Only 6.8×10^{-9} M Mg^{2+} will be in equilibrium with a saturated $Ca(OH)_2$ solution. This represents only $(6.8 \times 10^{-9}/5.5 \times 10^{-2}) = 1.3 \times 10^{-7}$ of the original, so the fraction of magnesium that can be removed by this method is $1 - (1.3 \times 10^{-7}) = $ **0.99999987** (i.e. virtually 100%).

10.77 If the relative humidity is 20% and the vapor pressure of water at 25°C is 23.8 torr, the pressure of water vapor is $(23.8$ torr$)(0.20) = 4.76$ torr. Since the vapor pressure of the $CuSO_4$ system is 5.60 torr, it will give up water to the atmosphere to establish equilibrium.

10.79 The dissolving of an ionic substance in water involves at least two phenomena: first, the actual dissolving of the solid to give free ions, and second, the reaction of these ions with water molecules. The latter involves interaction of the charged ions with the dipole of the water molecule, and typically involves the release of heat. Some ions are only weakly hydrated, and the energy released by the hydration is small compared to that of the dissolution. For a few ions, however, the ion-dipole interaction is very strong and releases a large amount of heat. Among the ions for which this is true are the hydronium and hydroxide ions. The heat released when acids or bases dissolve in water is largely due to the hydration of their ions. As a consequence, the hydration energy masks the effect of any energy absorbed or released by the dissolving itself. Thus, strong acids and bases liberate large amounts of heat energy when they are dissolved in water, and the temperature of the solution increases dramatically.

CHAPTER ELEVEN

Acids and Bases

11.1 Use $K_w = [H_3O^+][OH^-] = 1.0 \times 10^{-14}$. Then
$[OH^-] = 1.0 \times 10^{-14}/5.7 \times 10^{-8} = \mathbf{1.8 \times 10^{-7}\ M}$.

11.3 In order to perform these conversions, we use the equation $pH = -\log[H_3O^+]$.
   a. $pH = -\log(1.3 \times 10^{-2}) = \mathbf{1.89}$
   b. $pH = -\log(0.0231) = \mathbf{1.636}$
   c. $pH = -\log(7.9 \times 10^{-8}) = \mathbf{7.10}$

11.5 In order to perform these calculations, we will use the equations
$[H_3O^+] = 10^{-pH}$ and $[OH^-] = (1.0 \times 10^{-14})/[H_3O^+]$.
   a. $[H_3O^+] = 10^{-1.2} = \mathbf{0.06\ M}$
   $[OH^-] = \mathbf{2 \times 10^{-13}\ M}$
   b. $[H_3O^+] = 10^{-6.7} = \mathbf{2 \times 10^{-7}\ M}$
   $[OH^-] = \mathbf{5 \times 10^{-8}\ M}$
   c. $[H_3O^+] = 10^{-13.4} = \mathbf{4 \times 10^{-14}\ M}$
   $[OH^-] = \mathbf{0.25\ M}$

11.7 a. Arrhenius **acid**
   $HCl + H_2O \rightarrow \mathbf{H_3O^+} + Cl^-$
   b. Neither
   $NaCl \rightarrow Na^+ + Cl^-$
   c. Arrhenius **acid**
   $NH_4Cl \rightarrow NH_4^+ + Cl^-$
   $NH_4^+ + H_2O \rightarrow NH_3 + \mathbf{H_3O^+}$

11.9 a. Bronsted acid
   $NaHSO_4 \rightarrow Na^+ + HSO_4^-$
   $HSO_4^- + H_2O \rightarrow H_3O^+ + SO_4^{2-}$
   b. Bronsted base
   $Ca_3(PO_4)_2 \rightarrow 3Ca^{2+} + 2PO_4^{3-}$
   $PO_4^{3-} + H_2O \rightarrow HPO_4^{2-} + OH^-$
   c. Bronsted acid
   $HOCl + H_2O \rightarrow H_3O^+ + OCl^-$

CHAPTER ELEVEN

11.11 a. HCN loses a proton to form its conjugate base, **CN⁻**.
b. HSO_4^- loses a proton to form its conjugate base, **SO_4^{2-}**.
c. NH_4^+ loses a proton to form its conjugate base, **NH_3**.

11.13 a. Since HNO_2 is a weaker acid than HNO_3, **NO_2^- is a stronger base** than NO_3^-.
b. Since HCO_3^- is a weaker acid than H_2CO_3, **CO_3^{2-} is a stronger base** than HCO_3^-.
c. Since $H_2SO_3^-$ is a weaker acid than H_2SO_4, **HSO_3^- is a stronger base** than HSO_4^-.

11.15 a. $K = 1/K_a(NH_4^+) = K_b(NH_3)/K_w$
$K = 1.8 \times 10^{-5}/1.0 \times 10^{-14} = 1.8 \times 10^9$

The **forward** direction is favored because $K > 1$.
b. $K = K_w/K_a(HCN)$
$K = 1.0 \times 10^{-14}/7.2 \times 10^{-10} = 1.4 \times 10^{-5}$

The **reverse** direction is favored because $K < 1$.

11.17 HCl is a strong acid in aqueous solution and dissociates completely to H_3O^+ and Cl^- ions. In this case,
$[H_3O^+] = 0.0010$ M and $[OH^-] = (1.0 \times 10^{-14})/(0.0010) = $ **1.0×10^{-11} M**

11.19 a. For HOCl, $K_a = 3.7 \times 10^{-8}$.
$HOCl + H_2O \rightarrow H_3O^+ + OCl^-$
$[H_3O^+] = [OCl^-] = x$, $[HOCl] = 0.10$ M $- x$
$3.7 \times 10^{-8} = x^2/(0.10 - x) \approx x^2/(0.10)$
$x \approx 6.1 \times 10^{-5}$ M
Check: $0.10 - 6.1 \times 10^{-5} = 0.10$ so the approximation is valid.
$[H_3O^+] = 6.1 \times 10^{-5}$ M and $[OH^-] = 1.0 \times 10^{-14}/[H_3O^+] = $ **1.6×10^{-10} M**
$pH = -\log[H_3O^+] = -\log(6.1 \times 10^{-5}) = $ **4.21**
b. For HNO_2, $K_a = 4.5 \times 10^{-4}$.
$HNO_2 + H_2O \rightarrow H_3O^+ + NO_2^-$
$[H_3O^+] = [NO_2^-] = x$, $[HNO_2] = 0.10 - x$
$4.5 \times 10^{-4} = x^2/(0.10 - x) \approx x^2/(0.10)$ so $x \approx 6.7 \times 10^{-3}$ M
Check: 6.7×10^{-3} is about 7% of 0.10 M, so the approximation is marginally valid.
$[H_3O^+] = 6.7 \times 10^{-3}$ M and **$[OH^-] = 1.5 \times 10^{-12}$ M**
$pH = -\log(6.7 \times 10^{-3}) = $ **2.17**

CHAPTER ELEVEN

11.21 a. KOH is a strong base so **[OH$^-$] = 0.15 M.**
[H$_3$O$^+$] = (1.0 × 10^{-14})/[OH$^-$] = **6.7 × 10^{-14} M**
pH = −log(6.7 × 10^{-14}) = **13.17**

b. Methylamine has K$_b$ = 4.4 × 10^{-4}.
If we represent methylamine with the symbol B, then
B + H$_2$O → BH$^+$ + OH$^-$
[B] = 0.15 − x, [BH$^+$] = [OH$^-$] = x
4.4 × 10^{-4} = x^2/(0.15 − x) ≈ x^2/0.15
x ≈ 8.1 × 10^{-3} (about 5% change - valid approximation)
[OH$^-$] = 8.1 × 10^{-3} M; [H$_3$O$^+$] = 1.0 × 10^{-14}/[OH$^-$] = **1.2 × 10^{-12} M**
pH = −log(1.2 × 10^{-12}) = **11.92**

11.23 K$_a$ for HCN = 7.2 × 10^{-10}
(1.50 g HCN)(1 mol/27.02 g) = 5.55 × 10^{-2} moles HCN
(0.300 g NaCN)(1 mol/49.01 g) = 6.12 × 10^{-3} moles NaCN
(5.55 × 10^{-2} mol)/(0.750 L) = 0.0740 M HCN
(6.12 × 10^{-3} mol)/(0.750 L) = 0.00816 M CN$^-$
HCN + H$_2$O → H$_3$O$^+$ + CN$^-$
7.2 × 10^{-10} = [H$_3$O$^+$][CN$^-$]/[HCN] = x(0.00816 + x)/(0.0740 − x)
7.2 × 10^{-10} ≈ x(0.00816)/(0.0740), so x ≈ **6.5 × 10^{-9} M = [H$_3$O$^+$]**
[OH$^-$] = 1.0 × 10^{-14}/[H$_3$O$^+$] = **1.5 × 10^{-6} M**

11.25 HNO$_2$ + H$_2$O → H$_3$O$^+$ + NO$_2^-$
K$_a$ = 4.5 × 10^{-4} = [H$_3$O$^+$][NO$_2^-$]/[HNO$_3$]
4.5 × 10^{-4} = x^2/(0.10 − x) ≈ x^2/0.10
x ≈ 6.7 × 10^{-3} (6.7% change, approximation valid)

Fraction of dissociation = (amount dissociated)/(initial amount)
= 6.7 × 10^{-3}/0.10 = 6.7 × 10^{-2} = **6.7%**

11.27 First, we will calculate the original [H$_3$O$^+$] in 0.100 M HCN.
K$_a$ = 7.2 × 10^{-10} = [H$_3$O$^+$][CN$^-$]/[HCN]
7.2 × 10^{-10} = x^2/(0.100 − x) ≈ x^2/(0.100)
x^2 ≈ (7.2 × 10^{-10})(0.100) = 7.2 × 10^{-11}, x ≈ 8.5 × 10^{-6}
Check: (0.100 − 8.5 × 10^{-6}) = 0.100 so the approximation is valid.
[H$_3$O$^+$] = x = 8.5 × 10^{-6} M

CHAPTER ELEVEN

Next, we will calculate the $[H_3O^+]$ in one liter of 0.100 M HCN with 0.100 g KCN:

(0.100 g KCN)(1 mol/65.12 g) = 1.54×10^{-3} mol KCN

(1.54×10^{-3} mol KCN)/(1.00 L) = 1.54×10^{-3} M CN⁻

$K_a = 7.2 \times 10^{-10} = (x)(1.54 \times 10^{-3} + x)/(0.100 - x)$

$7.2 \times 10^{-10} \approx (x)(1.54 \times 10^{-3})/(0.100)$

$x \approx 4.7 \times 10^{-8}$ $x = 4.7 \times 10^{-8}$

Check: $(0.100 - 4.7 \times 10^{-8}) = 0.100$ so that the approximation is valid

$(1.54 \times 10^{-3} + 4.7 \times 10^{-8}) = 1.54 \times 10^{-3}$ so both approximations are valid

$[H_3O^+] = x = 4.7 \times 10^{-8}$ M

The change in $[H_3O^+]$ is $(4.7 \times 10^{-8}) - (8.5 \times 10^{-6}) = -8.45 \times 10^{-6}$ M.

The $[H_3O^+]$ decreased by 8.45×10^{-6} M (or 99.4%).

11.29 We are starting with 0.50 M H_2SeO_3. Let x = extent of first dissociation. Let y = extent of second dissociation.

$K_{a1} = 3 \times 10^{-3} = [H_3O^+][HSeO_3^-]/[H_2SeO_3] = x^2/(0.50 - x)$

$K_{a2} = 5 \times 10^{-8} = [H_3O^+][SeO_3^{2-}]/[HSeO_3^-] = (x+y)y/(x-y)$

$x^2 = (3 \times 10^{-3})(0.50 - x)$; $x^2 + (3 \times 10^{-3})x - 1.5 \times 10^{-3} = 0$

$x = \frac{1}{2}\left[-3 \times 10^{-3} \pm \sqrt{(3 \times 10^{-3})^2 + 6 \times 10^{-3}}\right]$; x = 0.037 (and a negative root which will not yield a physically significant answer)

$0.037 + y^2 = (5 \times 10^{-8})(0.037 - y)$; $y^2 + 0.037y - 1.8 \times 10^{-9} = 0$

$y = \frac{1}{2}\left[-0.037 \pm \sqrt{(0.037)^2 + 7.2 \times 10^{-9}}\right]$; $y = 5 \times 10^{-8}$ (and a negative root which will not yield a physically significant answer)

$[H_2SeO_3] = (0.50 M - x) = $ **0.46 M**

$[HSeO_3^-] = (x - y) = $ **0.037 M**

$[SeO_3^{2-}] = y = $ **5×10^{-8} M**

$[H_3O^+] = (x + y) = $ **0.037 M**

11.31 a. $K = 1/K_b(HCOO^-) = K_a(HCOOH)/K_w$
   $K = 1.8 \times 10^{-5}/1.0 \times 10^{-14} = $ **1.8×10^9**

b. $K = K_b(CN^-) = K_w/K_a(HCN)$
   $K = 1.0 \times 10^{-14}/7.2 \times 10^{-10} = $ **1.4×10^{-5}**

c. $K = K_b(F^-) = K_w/K_a(HF)$
   $K = 1.0 \times 10^{-14}/6.8 \times 10^{-4} = $ **1.5×10^{-11}**

CHAPTER ELEVEN

11.33 First, we will use the pH to calculate the hydronium ion concentration.
$[H_3O^+] = 10^{-pH} = 10^{-2.37} =$ **4.3×10^{-3} M**

Now we can set up the equilibrium expression, letting x = [OAc$^-$].
$K_a = 1.8 \times 10^{-5} = (4.3 \times 10^{-3}$ M$) x/(1.0 - x)$
$1.8 \times 10^{-5} \approx (4.3 \times 10^{-3}) x; x \approx 4.2 \times 10^{-3}$
Check: $(1.0 - 4.3 \times 10^{-3}) = 1.0$ so the approximation is valid.
% ionization = $x/(1.0) = 4.3 \times 10^{-3} =$ **0.43%**

11.35 First, we will calculate the number of moles of the salt.
(1.000 g CH$_3$NH$_3$Cl)(1 mol)/67.518 g) = 0.01481 mol CH$_3$NH$_3$Cl
CH$_3$NH$_3$Cl dissociates completely to form CH$_3$NH$_3^+$ and Cl$^-$.
CH$_3$NH$_3^+$ + H$_2$O → CH$_3$NH$_2$ + H$_3$O$^+$
$K_a = K_w/K_b(CH_3NH_2) = 2.3 \times 10^{-11}$
If we let "x" moles of CH$_3$NH$_3^+$ hydrolyze into CH$_3$NH$_2$ and H$_3$O$^+$, then
[CH$_3$NH$_2$] = [H$_3$O$^+$] = x and [CH$_3$NH$_3^+$] = (0.01481 − x).
$2.3 \times 10^{-11} = x^2/(0.01481 - x) \approx x^2/(0.01481)$
$x^2 \approx 1.55 \times 10^{-9}, x \approx 3.9 \times 10^{-5}$
Check: $(0.01481 - 3.9 \times 10^{-5}) = 0.01477$ so the approximation is valid.
[H$_3$O$^+$] = 3.9×10^{-5} M; [OH$^-$] = 2.5×10^{-10} M
pH = 4.4

11.37 K_a and K_b for a conjugate pair are related by $K_a K_b = K_w = 1.0 \times 10^{-14}$ and $K_a = 3.7 \times 10^{-8}$ for HOCl. Therefore
$K_b = 1.0 \times 10^{-14}/3.7 \times 10^{-8} =$ **2.7×10^{-7}**

11.39 We will start by calculating the molar concentration of NaF in the river water.
(0.137 g NaF)(1 mol/41.99 g) = 3.26×10^{-3} mol NaF per liter of solution
F$^-$ + H$_2$O → HF + OH$^-$
$K_b = K_w/K_a(HF) = 1.5 \times 10^{-11} = [HF][OH^-]/[F^-]$
We will let "x" moles per liter of F$^-$ hydrolyze to form HF and OH$^-$.
$1.5 \times 10^{-11} = x^2/(0.00326 - x) \approx x^2/0.00326)$
$x^2 \approx 4.89 \times 10^{-14}, x \approx 2.2 \times 10^{-7}$
Check: $(0.00326 - 2.2 \times 10^{-7}) = 0.00326$ so the approximation is valid.
[OH$^-$] = 2.2×10^{-7} M; [H$_3$O$^+$] = 4.5×10^{-8} M; pH = **7.35**

CHAPTER ELEVEN

11.41 We will start by calculating the concentration of HCl in the diluted solution.
$c_f = c_i V_i / V_f$ = (25.0 mL)(0.250 M)/(33.3 mL) = 0.188 M
Since HCl is a strong acid, $[H_3O^+]$ = 0.188 M. **pH = 0.726**

In the initial solution, the concentration of HCl is 0.250 M. Since HCl is a strong acid, $[H_3O^+]$ = 0.250 M. **pH = 0.602**

11.43 $HOAc + H_2O \rightarrow H_3O^+ + OAc^-$, $K_a = 1.8 \times 10^{-5}$
If we let "x" moles per liter of acetic acid ionize, then
$[HOAc] = (0.10 - x)$, $[OAc^-] = (0.10 + x)$ and $[H_3O^+] = x$.
$1.8 \times 10^{-5} = x(0.10 + x)/(0.10 - x) \approx x(0.10/0.10)$
$x \approx 1.8 \times 10^{-5}$. Check: $(0.10 \pm 1.8 \times 10^{-5}) = 0.10$ so the approximation is valid.
$[H_3O^+] = 1.8 \times 10^{-5}$ M; **pH = 4.74**

11.45 $HOAc + H_2O \rightarrow H_3O^+ + OAc^-$
$K_a = 1.8 \times 10^{-5} = [H_3O^+][OAc^-]/[HOAc]$
In order to have pH = 7.5, $[H_3O^+] = 10^{-7.5} = 3 \times 10^{-8}$ M.
$[OAc^-]/[HOAc] = K_a/[H_3O^+] = (1.8 \times 10^{-5})/(3 \times 10^{-8}) = 600$
[HOAc]/[OAc⁻] = 0.002
This would be an extremely ineffective buffer. The buffer component concentrations are well beyond a factor of ten of each other.

11.47 $HCN + H_2O \rightarrow H_3O^+ + CN^-$
K_a for HCN = 7.2×10^{-10}

a. If we let "x" moles per liter of HCN ionize, then
$[HCN] = (0.10 - x)$, $[CN^-] = (0.12 + x)$ and $[H_3O^+] = x$.
$7.2 \times 10^{-10} = x(0.12 + x)/(0.10 - x) \approx x(0.12/0.10)$
$x \approx 6.0 \times 10^{-10}$ (approximation valid)
$[H_3O^+] = 6.0 \times 10^{-10}$ M; **pH = 9.22**

b. If 0.010 mol HCl is added to the buffer, then the conjugate base (CN⁻) will absorb the excess acid.
$H_3O^+ + CN^- \rightarrow HCN + H_2O$ to produce $[CN^-]$ = 0.11 M, $[HCN]$ = 0.11 M
If we now let "x" moles per liter of HCN ionize in order to establish equilibrium, then $[HCN] = (0.11 - x)$, $[CN^-] = (0.11 + x)$ and $[H_3O^+] = x$.
$7.2 \times 10^{-10} = x(0.11 + x)/(0.11 - x) \approx x(0.11/0.11)$
$x \approx 7.2 \times 10^{-10}$ (approximation valid)
$[H_3O^+] = 7.2 \times 10^{-10}$ M; **pH = 9.14**

c. If 0.020 mol NaOH is added to the buffer, then the conjugate acid (HCN) will absorb the excess base.

$OH^- + HCN \rightarrow CN^- + H_2O$ to produce $[CN^-] = 0.13$ M, $[HCN] = 0.08$ M

If we now let "x" moles per liter of HCN ionize in order to establish equilibrium, then
$[HCN] = (0.08 - x)$, $[CN^-] = (0.13 + x)$ and $[H_3O^+] = x$.

$7.2 \times 10^{-10} = x(0.13 + x)/(0.08 - x)$

$7.2 \times 10^{-10} \approx x(0.13/0.08)$

$x \approx 4 \times 10^{-10}$ (approximation valid)

$[H_3O^+] = 4 \times 10^{-10}$ M; **pH = 9.4**

11.49 The balanced equation for the titration of an unknown base, B, with HCl is (assuming 1:1 stoichiometry):

$H_3O^+ + B \rightarrow BH^+ + H_2O$

At the equivalence point, the number of moles of added acid and base are equal:
$n_{acid} = n_{base} = (18.34$ mL$)(0.100$ M$) = 1.83$ mmol

$[B] = 1.83$ mmol$/25.00$ mL $= \mathbf{7.32 \times 10^{-2}}$ **M**

11.51 The balanced equation for the titration of an unknown, monoprotic acid, HA, with NaOH is:

$HA + OH^- \rightarrow A^- + H_2O$

At the equivalence point, the number of moles of added acid and base are equal:
$n_{acid} = n_{base} = (40.0$ mL$)(0.0760$ M$) = 3.04$ mmol.

We can calculate the molar mass of the acid from its mass and mole information:
$M = m/n. = 243.2$ mg$/3.04$ mmol $= \mathbf{80.0}$ **g/mol**.

11.53 pH = 6.7; $[H^+] = 10^{-6.7} = 2.0 \times 10^{-7}$ M Halfway to the equivalence point, half the acid HA has been converted to its conjugate base A^-, so $[HA] = [A^-]$

$K_a = [A^-][H^+]/[HA] = [H^+]$

$= \mathbf{2.0 \times 10^{-7}}$

11.55 The balanced equation for the titration of unknown, monoprotic acid, HA, with NaOH is:

$OH^- + HA \rightarrow H_2O + A^-$

At the equivalence point, the number of moles of added acid and base are equal.
$n_{base} = n_{acid} = (28.2$ mL$)(0.100$ M$) = 2.82$ mmol

CHAPTER ELEVEN

When 19.2 mL of base had been added to the original quantity of acid, n_{HA} = 2.82 mmol, and n_{OH^-} = (19.2 mL)(0.10 M) = 1.92 mmol; these react according to the equation given above to produce 1.92 mmol A^-, with 0.90 mmol HA remaining.

$HA + H_2O \rightarrow H_3O^+ + A^-$

$K_a = [H_3O^+][A^-]/[HA] = (10^{-7.76})(1.92 \text{ mmol}/V_{tot})/(0.90 \text{ mmol}/V_{tot})$

$K_a = (1.74 \times 10^{-8})(1.92/0.90) = \mathbf{3.7 \times 10^{-8}}$

11.57 In order to perform these calculations, we will use $[H_3O^+] = 10^{-pH}$.
   a. $[H_3O^+] = 10^{-2.97} = \mathbf{1.1 \times 10^{-3} \text{ M}}$
   b. $[H_3O^+] = 10^{-8.31} = \mathbf{4.9 \times 10^{-9} \text{ M}}$
   c. $[H_3O^+] = 10^{-0.04} = \mathbf{0.91 \text{ M}}$

11.59 K_a for acetic acid (HA) is 1.8×10^{-5}. Then $K_a = [H_3O^+][A^-]/[HA]$
   a. Let x be the equilibrium concentration of A^- or H_3O^+. Then the equilibrium expression becomes
   $K_a = 1.8 \times 10^5 = x^2/1.0 - x$. We neglect x compared to 1.0, giving
   $1.8 \times 10^{-5} = x^2/1.0$
   $x = 4.2 \times 10^{-3} \text{ M} = \mathbf{[H_3O^+] = [A^-]}$
   [HA] remains 1.0 M.
   b. Now we have
   $1.8 \times 10^{-5} = x^2/(0.010 - x)$. Again, neglect x in the denominator:
   $1.8 \times 10^{-5} = x^2/0.010$
   $x = \mathbf{4.2 \times 10^{-4} \text{ M} = [H_3O^+] = [A^-]}$
   and **[HA] = 0.010 M.**

11.61 The stronger the acid, the weaker the conjugate base. From Table 11.2, we see that acid strengths are in the order $HNO_2 > HF > HOCl > HCN$, so the conjugate base strength order would be $NO_2^- < F^- < OCl^- < CN^-$.
   K_b for H_2O is 1×10^{-14}, so it is weaker than all of these bases. K_b for NH_3 is 1.8×10^{-5}, so its $K_a = K_w/K_b = 1 \times 10^{-14}/1.8 \times 10^{-5} = 5.6 \times 10^{-10}$, lower than K_a for HCN. Therefore, it is a stronger base than CN^-.
   Even in NH_3, OH^- still acts as a base, so it is a stronger base than NH_3. It is much more difficult to remove a proton from NH_3 to form NH_2^- than it is to remove a proton from H_2O to form OH^-, so NH_2^- is the strongest base of all. Ammonia is a stronger base than water, so the final order of increasing basicity is
   $H_2O < NO_2^- < F^- < OCl^- < CN^- < NH_3 < OH^- < NH_2^-$

CHAPTER ELEVEN

11.63 The two problems deal with 1.0 M and 0.10 M acetic acid solutions, respectively. The H_3O^+ concentration drops, but not nearly by a factor of ten. The two equilibrium expressions were

$1.8 \times 10^{-5} = x^2/1.0$ and

$1.8 \times 10^{-5} = x^2/0.10$

where the x in the denominators has been neglected. We see that x will be proportional to the square root of the acid concentration. Thus, a 10-fold change in [HA] will result in $[H_3O^+]$ changing by only a factor of about three.

11.65 (0.125 mol HF)/(0.200 L) = 0.625 M HF
(0.250 mol F^-)/(0.200 L) = 1.25 M F^-
$HF + H_2O \rightarrow H_3O^+ + F^-$
$K_a = 6.8 \times 10^{-4} = [H_3O^+][F^-]/[HF]$
$3.53 \times 10^{-4} = x(1.25 + x)/(0.625 - x) \approx x(1.25)/(0.625)$
$x = 3.4 \times 10^{-4}$ (approximation valid)
$[H_3O^+] = 3.4 \times 10^{-4}$ M, and **pH = 3.47**

11.67 a. pH = 7.00
   b. pH = 7.00
   c. $NH_3 + H_2O \rightarrow NH_4^+ + OH^-$
      $K_b = 1.8 \times 10^{-5} = [NH_4^+][OH^-]/[NH_3]$
      $1.8 \times 10^{-5} = x^2/(0.10 - x) \approx x^2/(0.10)$
      $x \approx 1.3 \times 10^{-3}$ (approximation valid)
      $[OH^-] = 1.3 \times 10^{-3}$ M so $[H_3O^+] = 7.7 \times 10^{-12}$ M
      pH = 11.11
   d. $HOAc + H_2O \rightarrow OAc^- + H_3O^+$
      $K_a = 1.8 \times 10^{-5} = [H_3O^+][OAc^-]/[HOAc]$
      $1.8 \times 10^{-5} = x^2/(0.10 - x) \approx x^2/(0.10)$
      $x \approx 1.3 \times 10^{-3}$ (approximation valid)
      $[H_3O^+] = 1.3 \times 10^{-3}$ M so **pH = 2.89**
   e. $NH_4^+ + H_2O \rightarrow NH_3 + H_3O^+$
      $K_a = 5.6 \times 10^{-10} = [H_3O^+][NH_3]/[NH_4^+]$
      $5.6 \times 10^{-10} = x^2/(0.10 - x) \approx x^2/(0.10)$
      $x = 7.5 \times 10^{-6}$ (approximation valid)
      $[H_3O^+] = 7.5 \times 10^{-6}$ M so **pH = 5.12**

CHAPTER ELEVEN

    f. $OAc^- + H_2O \rightarrow HOAc + OH^-$
$K_b = 5.6 \times 10^{-10} = [HOAc][OH^-]/[OAc^-]$
$5.6 \times 10^{-10} = x^2/(0.10 - x) \approx x^2/(0.10)$
$x = 7.5 \times 10^{-6}$ (approximation valid)
$[OH^-] = 7.5 \times 10^{-6}$ M so $[H_3O^+] = 1.3 \times 10^{-9}$ M.
pH = 8.89

    g. $NH_4^+ + H_2O \rightarrow NH_3 + H_3O^+$
$K_a = 5.6 \times 10^{-10} = [H_3O^+][NH_3]/[NH_4^+]$
$5.6 \times 10^{-10} = x(0.10 + x)/(0.10 - x) \approx x(0.10)/(0.10)$
$x = 5.6 \times 10^{-10}$ (approximation valid)
$[H_3O^+] = 5.6 \times 10^{-10}$ M, and **pH = 9.25**

11.69 $HOCl + H_2O \rightarrow OCl^- + H_3O^+$
$K_a = 3.7 \times 10^{-8}$
If we let "x" moles of HOCl ionize into OCl^- and H_3O^+, then $[OCl^-] = [H_3O^+] =$ x and $[HOCl] = (0.050 - x)$.
$3.7 \times 10^{-8} = x^2/(0.050 - x) \approx x^2/(0.050)$
$x \approx 4.3 \times 10^{-5}$ (approximation valid)
$[H_3O^+] = 4.3 \times 10^{-5}$ M; $[OH^-] = 2.3 \times 10^{-10}$ M
pH = 4.37

11.71 a. In water, HCl(aq) is essentially completely dissociated into H_3O^+ and Cl^- ions. H_3O^+ reacts with the base NH_3:
$NH_3 + H_3O^+ + Cl^- \rightarrow NH_4^+ + Cl^-$

      The net ionic reaction remains when Cl^- is eliminated from both sides of the equation:
$NH_3 + H_3O^+ \rightarrow NH_4^+ + H_2O$

    b. Similarly, HNO_3 forms H_3O^+ and NO_3^- ions in water:
$(CH_3)_3N + H_3O^+ \rightarrow (CH_3)_3NH^+ + H_2O$

    c. HF in contrast is weakly acidic, so the reaction is
$(CH_3)_2NH + HF \rightarrow (CH_3)_2NH_2^+ + F^-$

    d. The first ionization step of H_2SO_4 (to HSO_4^- and H_3O^+) is strong, and the second is moderately weak. Therefore, the net ionic reaction is:
$H_3O^+ + HSO_4^- + 2(CH_3)_2NH \rightarrow 2(CH_3)_2NH_2^+ + SO_4^{2-} + H_2O$

CHAPTER ELEVEN

11.73 For neutral water at any temperature, $[H_3O^+] = [OH^-]$, and $K_w = [H_3O^+][OH^-]$. At 0°C, $K_w = 1.1 \times 10^{-15} = x \cdot x = x^2$ where x is the concentration of either of the ions. Then
$1.1 \times 10^{-15} = x^2$; $x^2 = 3.3 \times 10^{-7}$ M
pH = log (3.3×10^{-7}) = **6.48**

11.75 a. Applying Henry's law to CO_2, we get $p_{CO_2} = K_H \cdot X_{CO_2}$.
1.00 atm = (1640 atm) $[n_{CO_2}/n_{H_2O}]$
n_{CO_2} = (1.00 atm)(55.55 mol)/(1640 atm) = 0.0339 mol
(0.0339 mol)/(1.00 L) = **0.0339 M CO_2**

b. We are assuming that 0.0339 M CO_2 produces 0.0339 M H_2CO_3. Since $K_{a1} = 4.3 \times 10^{-7}$ and $K_{a2} = 4.4 \times 10^{-11}$ for CO_2, we will assume that the $[H_3O^+]$ generated in the second dissociation is negligible in comparison with the $[H_3O^+]$ generated in the first dissociation. So the pH is determined by the first dissociation only. If we let "x" moles per liter of H_2CO_3 dissociate, then $[H_3O^+] = [HCO_3^-]$ = x and $[H_2CO_3]$ = (0.0339 – x).
$4.3 \times 10^{-7} = x^2/(0.0339 - x) \approx x^2/(0.0339)$
x ≈ 1.2×10^{-4} (approximation valid)
$[H_3O^+]$ = x = 1.2×10^{-4} M; **pH = 3.92**

11.77 The freezing point depression constant for water is K_f = –1.86 K/m (Table 9.5).
Since the density of the solution is 1.000 g/mL, then 1 kg solvent occupies 1 L. Thus, a 0.2000 m solution is also 0.2000 M. To solve this problem, we will first determine the concentration of solute particles from the freezing point depression:
molality = $\Delta T/K_f$ = –0.3785°/(–1.86 deg/m)
= 0.2035 m (we will carry an extra digit throughout)
= 0.2035 M
This is larger than the given concentration because of the partial dissociation of benzoic acid into ions:
H_2O + HB → H_3O^+ + B^-
0.2000 – x   x   x
where x is the concentration of the benzoic acid that has formed ions. Then the total concentration of particles is 0.2035 M = 0.2000 – x + x + x = 0.2000 + x
Therefore x = 0.0035 M.
The fraction of ionization is 0.0035/0.2000 = 0.018 or **1.8%**
$K_a = [H_3O^+][C_6H_5COO^-]/[C_6H_5COOH] = (0.0035)^2/(0.1965)$
$K_a = 6.2 \times 10^{-5}$

CHAPTER ELEVEN

11.79 The number of moles of sulfur in the coal can be calculated as follows:
(0.0110)(1.00 × 10^3 g)(1 mol/32.06 g) = 0.343 mol S
This quantity of sulfur could be converted to 0.343 mol H$_2$S, 0.343 mol SO$_2$ or 0.343 mol SO$_3$.

H$_2$S:
(0.343 mol H$_2$S)/(10.0 L) = 0.0343 M H$_2$S
Since K$_{a1}$ >> K$_{a2}$, we will assume that [H$_3$O$^+$] produced in the second dissociation is negligible in comparison with [H$_3$O$^+$] produced in the first dissociation. If we let "x" moles per liter of H$_2$S dissociate, then
[H$_2$S] = (0.0343 − x), [H$_3$O$^+$] = [HS$^-$] = x.
K$_{a1}$ = 9.1 × 10^{-8} = x^2/(0.0343 − x) ≈ x^2/(0.0343)
x ≈ 5.6 × 10^{-5} (approximation valid)
[H$_3$O$^+$] = x = 5.6 × 10^{-5} M; **pH = 4.25**

SO$_2$:
SO$_2$ reacts with water to form H$_2$SO$_3$.
(0.343 mol H$_2$SO$_3$)/(10.0 L) = 0.0343 M H$_2$SO$_3$
Since K$_{a1}$ >> K$_{a2}$, we will assume that [H$_3$O$^+$] produced in the second dissociation is negligible in comparison with [H$_3$O$^+$] produced in the first dissociation. If we let "x" moles per liter of H$_2$SO$_3$ dissociate, then
[H$_2$SO$_3$] = (0.0343 − x), [H$_3$O$^+$] = [HSO$_3^-$] = x.
K$_{a1}$ = 1.7 × 10^{-2} = x^2/(0.0343 − x)
Since K$_{a1}$ is so large, we will solve this equation by using the quadratic equation rather than by assuming "x" is small.
5.83 × 10^{-4} − 0.017x = x^2
x^2 + (0.017)x − (5.83 × 10^{-4}) = 0
$$x = \frac{1}{2}\left[-0.017 \pm \sqrt{2.89 \times 10^{-4} + 2.332 \times 10^{-3}}\right]$$; x = 0.017 (plus a negative root which will not yield a physically reasonable answer)
[H$_3$O$^+$] = x = 0.017 M; **pH = 1.77**

SO$_3$:
SO$_3$ reacts with water to form H$_2$SO$_4$.
(0.343 mol H$_2$SO$_4$)/(10.0 L) = 0.0343 M H$_2$SO$_4$
The first dissociation of H$_2$SO$_4$ is a strong acid dissociation which will produce 0.0343M H$_3$O$^+$ and 0.0343 M HSO$_4^-$. The second dissociation has K$_{a2}$ = 2 × 10^{-2}.
If we let "x" moles per liter of HSO$_4^-$ dissociate, then [HSO$_4^-$] = (0.0343 − x),
[H$_3$O$^+$] = (0.0343 + x) and [SO$_4^{2-}$] = x.
K$_{a2}$ = 0.02 = x(0.0343 + x)/(0.0343 − x)

CHAPTER ELEVEN

Since K_{a2} is so large, we will solve this equation by using the quadratic equation rather than by assuming "x" is small.
$6.86 \times 10^{-4} - 0.02x = 0.0343x + x^2$
$x^2 + 0.0543x - 6.86 \times 10^{-4} = 0$

$x = \frac{1}{2}\left[-0.0543 \pm \sqrt{2.95 \times 10^{-4} + 2.74 \times 10^{-3}}\right]$; x = 0.0106 (plus a negative root which will not yield a physically reasonable answer).
$[H_3O^+] = (0.0343 + x) = 0.045$ M; **pH = 1.35**
SO_3 produces the most H_3O^+ (lowest pH) and would be most harmful to the environment.

11.81 We know that the number of moles of nitrogen in the coal sample will be equal to the number of moles of ammonia produced. The number of moles of ammonia can be calculated from the titration information.
moles N in coal sample = moles NH_3 produced
moles NH_3 = moles HCl to endpoint
moles NH_3 = (0.100 M)(0.0582 L) = 5.82×10^{-3} mol NH_3
So there was (5.82×10^{-3} mole N)(14.0067 g/mol) = 0.0815 g N in original sample
(0.0815 g N)/(9.33 g sample) = **0.874%** nitrogen in the coal sample.

11.85 K_{sp} for $PbSO_4$ = 1.3×10^{-8} = $[Pb^{2+}][SO_4^{2-}]$
K_a for HSO_4^- = 2.0×10^{-2} = $[H_3O^+][SO_4^{2-}]/[HSO_4^-]$
HCl dissociates completely, so $[H_3O^+]$ = 0.10 M.
All of the SO_4^{2-} and HSO_4^- in solution come from the $PbSO_4$, so
$[Pb^{2+}] = [SO_4^{2-}] + [HSO_4^-]$
The ratio of $[SO_4^{2-}]$ to $[HSO_4^-]$ can be determined as follows:
$K_a = 2.0 \times 10^{-2} = (0.10)[SO_4^{2-}]/[HSO_4^-]$
$[HSO_4^-] = (0.10)[SO_4^{2-}]/ 2.0 \times 10^{-2}$
$[HSO_4^-] = 5 \cdot [SO_4^{2-}]$ so, $[Pb^{2+}] = 6 \cdot [SO_4^{2-}]$
$1.3 \times 10^{-8} = [Pb^{2+}][SO_4^{2-}] = 6[SO_4^{2-}][SO_4^{2-}]$
$[SO_4^{2-}]^2 = (1.3 \times 10^{-8})/6 = 0.217 \times 10^{-8}$; $[SO_4^{2-}] = 4.65 \times 10^{-5}$ M.
From above, $[Pb^{2+}] = 6[SO_4^{2-}] = 6(4.65 \times 10^{-5}) = 2.8 \times 10^{-4}$ M. Since all the lead comes from the $PbSO_4$, the solubility of $PbSO_4$ in this solution is $\mathbf{2.8 \times 10^{-4}}$ **M**

CHAPTER ELEVEN

11.87 K_{sp} for CoS $= 4.0 \times 10^{-21} = [Co^{2+}][S^{2-}]$. In water, the solubility of CoS is $\sqrt{(4.0 \times 10^{-21})} = \mathbf{6.3 \times 10^{-11}\ M}$.

In 0.1 M H_2S at pH 2, the solution is 0.01 M in Co^{2+} and is 0.1 M in H_2S. First, determine the S^{2-} concentration from K_{a1}:

$K_{a1} = 9.1 \times 10^{-8} = [H_3O^+][HS^-]/[H_2S]$

Let $x = [HS^-]$

$9.1 \times 10^{-8} = x^2/(0.1 - x) \approx x^2/0.1$

$[HS^-] = 1 \times 10^{-4}$ M (0.1% change, approximation valid).

Next, determine the S^{2-} concentration from K_{a2}:

$K_{a2} = [H_3O^+][S^{2-}]/[HS^-] = 1.2 \times 10^{-15}$

Let $y = [S^-]$

$1.2 \times 10^{-15} = y^2/(1 \times 10^{-4} - y) \approx y^2/1 \times 10^{-4}$

$y = 4 \times 10^{-6}$ (4% change, approximation valid).

Since the starting Co^{2+} concentration was 0.01 M and the S^{2-} concentration is 4×10^{-6} M, the reaction quotient Q is $(0.01)(4 \times 10^{-6}) = 4 \times 10^{-8}$. This is much larger than K_{sp} for CoS, so **CoS will precipitate** from this solution.

11.89 $K_a(NH_4^+) = 5.6 \times 10^{-10} = [H_3O^+][NH_3]/[NH_4^+]$

$K_b(NH_3) = 1.8 \times 10^{-5} = [NH_4^+][OH^-]/[NH_3]$

Before any HCl is added, $V_{HCl} = 0.00$ mL and pH = 11.13.

At the endpoint, $V_{HCl} = 25.00$ mL and pH = 5.28.

Halfway to the endpoint, $V_{HCl} = 12.50$ mL and pH = 9.25.

When $V_{HCl} = 37.50$ mL, pH = 1.70.

When $V_{HCl} = 6.25$ mL, pH = 9.73.

When $V_{HCl} = 31.25$ mL, pH = 1.95.

A good indicator would be bromcresol green (with a color change of blue to yellow through the equivalence point) or methyl red (with a color change of yellow to red through the equivalence point).

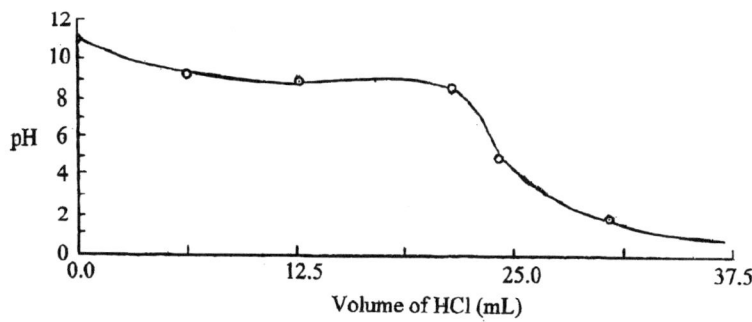

CHAPTER ELEVEN

11.91 a. At the equivalence point, the number of moles of added acid and base are equal. $n_{acid} = n_{base} = (42.60\text{ mL})(0.2500\text{ M}) = 10.65$ mmol
We can calculate the apparent molar mass of the acid from its mass and mole information:
$M = m/n = 2.344\text{ g}/0.01065\text{ mol} =$ **220.0 g/mol**

b. For each 100.00 mL of the acidic solution, (HA) = (2.344 g)(1 mol/220.0 g) = 0.01065 moles. [HA] = (0.01065 mol)/(0.1000 L) = 0.1065 M. At the endpoint, 100.00 mL of 0.1065 M HA has reacted with 42.60 mL of 0.2500 M OH⁻ to form 142.60 mL of (0.1065 M)(100.0 mL/142.60 mL) = 0.0747 M A⁻.
$A^- + H_2O \rightarrow HA + OH^-$
$K_b = [HA][OH^-]/[A^-] = x^2/(0.0747 - x)$

But "x" is known from the pH at the endpoint.
pH = 9.40; pOH = 4.60; $[OH^-] = 10^{-4.60} = 2.5 \times 10^{-5} = x$
$K_b = (2.5 \times 10^{-5})^2/(0.0747) = 8.4 \times 10^{-9}$
$K_a = K_w/K_b =$ **1.2×10^{-6}**

c. Initially, [HA] = 0.1065 M.
$HA + H_2O \rightarrow H_3O^+ + A^-$
$K_a = [H_3O^+][A^-]/[HA] = x^2/(0.1065 - x) \approx x^2/0.1065$
$1.2 \times 10^{-6} \approx x^2/0.1065$
$x \approx 3.6 \times 10^{-4}$ (approximation valid--0.34% error)
$[H_3O^+] = x = 3.6 \times 10^{-4}$ so **pH = 3.44**

d. At the midpoint of titration, $[A^-] = [HA]$ so $[H_3O^+] = K_a$.
$[H_3O^+] = 1.2 \times 10^{-6}$ M; **pH = 5.92**

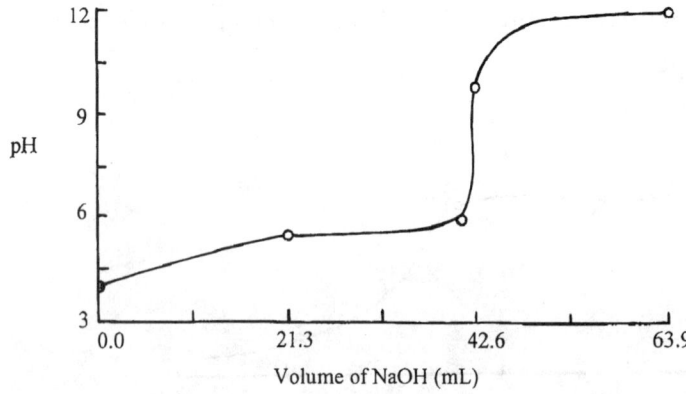
Volume of NaOH (mL)

CHAPTER TWELVE

Heat, Work, and Energy

12.1 Use $q = n \cdot C \cdot \Delta T$, where C is the heat capacity.
   a. $C_{H_2O} = 75.4$ J/mol·K from Table 12.2.
      $q = (75.4 \text{ J/mol·K})(1.50 \times 10^3 \text{ g}/18.01 \text{ g/mol})(60° - 30°) = 1.9 \times 10^5$ **J**
   b. $C_{Fe} = 25.1$ J/mol·K
      $q = (25.1 \text{ J/mol·K})(2.45 \text{ g}/55.85 \text{ g/mol})(28° - 25°) = $ **3 J**

12.3 Since the internal energy *drops*, $\Delta E = -500.$ J. Since work is done *on* the system, $w = +150$ J. Now, the heat can be calculated from the relationship:
   $q = \Delta E - w$.
   $q = (-500. \text{ J}) - (150. \text{ J}) = $ **−650. J**.
   Based on the sign of q, we can say that 650. J of heat were transferred *from* the system to the surroundings.

12.5 We know that $q = 2.00$ kJ. We can calculate the work from $w = -P\Delta V$.
   $w = -(5.0 \text{ atm})(20.0 \text{ L} - 10.0 \text{ L}) = -50 \text{ L·atm} = -5066 \text{ J} = -5.1 \text{ kJ}$
   $\Delta E = q + w = 2.00 \text{ kJ} - 5.1 \text{ kJ} = $ **−3.1 kJ**

12.7 a. First, we will calculate the number of moles of gas present by $n = PV/RT$:
      $(10 \text{ atm})(1 \text{ L})/[(0.08206 \text{ L·atm/mol·K})(273 \text{ K})] = 0.45$ moles of gas
      Since this is a constant temperature process, $\mathbf{\Delta E = 0}$.
      Also, $\mathbf{\Delta H = 0}$ since $\Delta H = \Delta E + \Delta(PV) = 0 + \Delta(nRT) = 0$ (n, R and T are all constant).
      Since $\Delta E = 0$, $q = -w = P\Delta V$. We can calculate ΔV from knowledge of V_i and V_f.
      $V_i = 1$ L, $V_f = (nRT/P)_f = (0.45 \text{ mol})(0.08206 \text{ L·atm/mol·K})(273 \text{ K})/(1 \text{ atm})$
      $V_f = 10$ L
      So $q = (1 \text{ atm})(9 \text{ L}) = 9 \text{ L·atm} = $ **900 J**; $w = -q = $ **−900 J**
   b. If the expansion were an isothermal free expansion (that is, expansion against an external pressure of zero), work would be zero.
      Then, $-w = q = \mathbf{\Delta E = \Delta H = 0}$.

12.9 Calculate q and w and use $\Delta E = q + w$.
   $q = +500.$ J $= 0.500$ kJ
   $w = -P_{ex}\Delta V = -(1.500 \text{ atm})(20.00 \text{ L} - 10.0 \text{ L})$
      $= -115.0 \text{ L·atm} \times 101.3 \text{ J/L·atm} = -1.52$ kJ
   $\Delta E = q + w = 0.500 - 1.52 = $ **−1.02 kJ**

CHAPTER TWELVE

12.11   a. w = 0 if the volume is constant
        b. q = n·C·ΔT
           200. J = (2.0 mol)(12.4 J/mol·K)ΔT
           ΔT = 8.1° = $T_f - T_i$
           T_f = 25 + 8.1 = **33°C**

12.13 q_p = ΔH = –110. kJ
       q_v = ΔE. But ΔH and ΔE are related by ΔH = ΔE + Δn_{gas}RT
       Δn_{gas} = 1 – 1/2 = 1/2
       q_v = ΔE = ΔH – Δn_{gas}RT = –110. kJ – 1/2(8.314 J/mol·K)(10^{-3} kJ/J)(298)
       = –110. – 1.2 = **–111 kJ**

12.15   ΔH = q_p = 6.00 kJ
       ΔE = ΔH – PΔV
       ΔV = (1 mol H_2O)(18.02 g/mol)(1/1.00 – 1/0.915)
           = –1.67 mL = –1.67 × 10^{-3} L
       ΔE = 6.00 × 10^3 J – (–1.67 × 10^{-3} L)(1.00 atm)(101 L·atm/J)
       ΔE = 6.00 × 10^3 J

12.17   The process is K(l) → K(g). We are given that ΔE = 1530. kJ and that w = –216.2 kJ (negative because "work was done")
       a. ΔE and ΔH are related by
           ΔH = ΔE + PΔV
           w = –PΔV = –216.2 kJ
           ∴ ΔH = ΔE + 216.2 kJ
               = 1530. + 216.2 = 1746 kJ for 1.00 kg
               = (1.746 × 10^3 J/kg)(0.03910 kg/mol)
           ΔH = 68.27 kJ/mol
       b. From (a),
           PΔV = (216 kJ)/(101.3 J/L·atm)
               = 2130 L·atm
           Since we know that ΔV = 2134 L
           P = 2130 L·atm/2134 L
           P = P_{ex} = **1.00 atm**

CHAPTER TWELVE

12.19 a. We will write the equation for one mole of benzoic acid:
$C_6H_5CO_2H(s) + (7.5)O_2(g) \rightarrow 7CO_2(g) + 3H_2O(l)$

b. In a bomb calorimeter, there is no volume change during combustion, and **w = 0**.
q = –C ΔT where C refers to the calorimeter as a whole.
q = –(5.020 kJ/°C)(31.668 °C – 25.240 °C) = **–32.27 kJ**
ΔE = q + w = q = **–32.27 kJ**

c. In order to calculate ΔH for the reaction, we will first express ΔE for the reaction, on a "per-mole" basis. The formula mass for benzoic acid is 122.1 g/mol so the number of moles of benzoic acid combusted is 0.01000.
ΔE = –32.27 kJ/0.01000 mol = –3227 kJ/mol.
ΔH = ΔE + Δ(PV) = ΔE + (Δn_{gas})RT
ΔH = –3227 kJ/mol + (7 – 7.5)(8.314 × 10⁻³ kJ/mol·K)(298.39K)
ΔH = –3228 kJ/mol
(Note that in doing the ΔH calculation, we used the initial temperature in ΔnRT)

12.21 We can calculate the standard enthalpy change for this reaction from the standard enthalpies of formation for the reactants and products:
ΔH° = $\Delta H_f°$(C₂H₅OH(l)) – $\Delta H_f°$(H₂O(l)) – $\Delta H_f°$(C₂H₄(g))
ΔH° = –278 kJ/mol – (–286 kJ/mol) – (52.3 kJ/mol) = **–44.3 kJ/mol**

12.23 For each of the reactions, we can calculate the standard enthalpy change based on the standard enthalpies of formation for the reactants and products.
a. ΔH°₁ = $\Delta H_f°$(CO₂(g)) – $\Delta H_f°$(CO(g)) = (–393 + 110) kJ/mol
= **–283 kJ/mol**
b. ΔH°₂ = $\Delta H_f°$ (H₂O(l)) = **–286 kJ/mol**
c. ΔH°₃ = $\Delta H_f°$ (H₂O(g)) – $\Delta H_f°$(H₂O(l)) – (–242 + 286) kJ/mol = **44 kJ/mol**
d. Using Hess' law, we can write ΔH_rxn = ΔH°₁ – ΔH°₂ – ΔH°₃.
ΔH_rxn = **–41 kJ/mol**

12.25 We will assume that this is a constant pressure process so that heat = q_p = ΔH°.
ΔH° = $\Delta H_f°$(Al₂O₃(s)) – $\Delta H_f°$(Fe₂O₃(s))
ΔH° = –1676 kJ/mol + 822 kJ/mol = –854 kJ/mol
(–854 kJ/mol rxn)(1 mol rxn/2 mol Fe)(1 mol Fe/55.85 g Fe) = –7.64 kJ/g Fe
7.64 kJ of heat are released for each gram of metal produced.

12.27 ΔH° = 2$\Delta H_f°$(H₂O(l)) – 2 $\Delta H_f°$(H₂S(g)) – $\Delta H_f°$(SO₂(g))
ΔH° = 2(–286 kJ) – 2(–20.1 kJ) – (–297 kJ) = **–235 kJ**

CHAPTER TWELVE

12.29 First, we will write the balanced equation for the formation of $Ca(OH)_2$ from its elements: $Ca(s) + O_2(g) + H_2(g) \rightarrow Ca(OH)_2(s)$.
This equation can be generated by adding together the three equations given. Thus, $\Delta H_f°$ can be calculated by summing the $\Delta H°$ values.
$\Delta H_f°(Ca(OH)_2) = (-286 \text{ kJ}) + (-64 \text{ kJ}) + (-635 \text{ kJ}) = $ **–985 kJ**

12.31 The balanced equation is
$C_5H_{12}(l) + 8O_2(g) \rightarrow 5CO_2(g) + 6H_2O(l)$
From Table 12.3, the enthalpy of combustion $\Delta H_c°$ of pentane is –3487 kJ/mol.
$\Delta H°$ and $\Delta E°$ are related by $\Delta H° = \Delta E° + \Delta(PV) = \Delta E° + RT\Delta n_{gas}$.
$\Delta n_{gas} = 5 - 8 = -3$ Therefore,
$\Delta E° = \Delta H° - RT\Delta n_{gas}$
$= -3487 \text{ kJ/mol} - (8.314 \times 10^{-3} \text{ kJ/mol·K})(298 K)(-3 \text{ mol}) = -3480 \text{ kJ/mol}$
3480 kJ of heat are released upon combustion of one mole of n-pentane in a bomb calorimeter at room temperature.

12.33 Assume that the ignition and burning occur at constant pressure. We can calculate the heat released during the combustion (and added to the water) from the temperature change of the water.
$q = q_p = nC_p\Delta T = (100. \text{ g})(4.184 \text{ J/g °C})(22.2 °C) = 9288 \text{ J} = 9.29 \text{ kJ}$
This is the quantity of heat released when 1.00 g of sulfur is burned. We can express this on a "per-mole" basis if we calculate the number of moles of sulfur burned:
1.00 g sulfur (1 mol/32.06 g) = 0.0312 mol sulfur. We will express ΔH_c as a negative number since the combustion of sulfur is exothermic.
$\Delta H_c°(S) = (-9.29 \text{ kJ})/(0.0312 \text{ mol}) = $ **–298 kJ/mol**

12.35 First, write the balanced equation:
$3/2 \text{ } O_2(g) + (NH_2)_2CO(s) \rightarrow 2H_2O(l) + N_2(g)$
From Table 12.3 we see the heat of combustion of urea $\Delta H_c = -632$ kJ/mol. In order to compare this with the results from the bomb calorimeter, we will have to convert ΔH_c to ΔE. The relationship between them is
$\Delta E = \Delta H - (\Delta n_{gas})RT$
$\Delta n_{gas} = 1/2$ for this reaction
$\therefore \Delta E = -632 \text{ kJ/mol} - (8.3145 \text{ J/mol·K})(298K)(1/2)$
$= -633 \text{ kJ/mol}$.
1.370 g urea/60.06 g/mol = 2.281×10^{-2} mol;

CHAPTER TWELVE

(2.281 × 10⁻² mol)(–633 kJ/mol) = –14.4 kJ; 14.4 kJ of heat would be generated by the combustion if the substance was urea. Now analyze the calorimeter result. The heat absorbed by the calorimeter was (4.05 kJ/K)(27.70°C – 24.12°C) = 14.5 kJ

Thus, the experiment is **consistent with the substance being urea**.

12.37 18.02 g of water is 1.000 mol of water. The vaporization of the water requires 40.66 kJ of heat at constant pressure, q_p. Therefore, $\Delta H = q_p$ = 40.66 kJ for the vaporization. ΔH and ΔE are related by $\Delta H = \Delta E + \Delta n_{gas}RT$.
For $H_2O(l) \rightarrow H_2O(g)$, $\Delta n_{gas} = 1$ and so $\Delta n_{gas}RT = (1)(8.314 \text{ J/mol·K})(373K) = 3.10$ kJ.
Therefore $\Delta E = \Delta H - 3.10 = 40.66 - 3.10 =$ **37.56 kJ**

12.39 For the combustion of diborane, the heat of combustion can be written in terms of the heats of formation of products and reactants:
$\Delta H_c = \Delta H_f(B_2O_3) + 3\Delta H_f(H_2O) - \Delta H_f(B_2H_6)$.
We are given the heat of combustion (–1940 kJ) and the heat of formation of B_2O_3 (–2370 kJ). The value of ΔH_f for $H_2O(g)$ from Table 12.4 is –242 kJ/mol, so we can write for the combustion:
$\Delta H_c = -1940$ kJ $= -2370$ kJ $+ 3(-242$ kJ$) - \Delta H_f(B_2H_6)$.
Solving: $\Delta H_f(B_2H_6) =$ **1156 kJ/mol**.

12.41 For the reaction given,
$\Delta H° = 105$ kJ $= 2 \Delta H_f°(AgCl(s) - \Delta H_f°(PbCl_2(s)))$.
We can calculate the heat of formation of $PbCl_2(s)$ since we can find the heat of formation for $AgCl(s)$ in Table 12.4.
105 kJ = 2(–127 kJ) – $\Delta H_f°(PbCl_2(s))$; $\Delta H_f°(PbCl_2(s)) =$ **–359 kJ**

12.43 The equation for the formation of ethylene from acetylene is as follows:
$C_2H_2(g) + H_2(g) \rightarrow C_2H_4(g)$
This equation can be generated by subtracting the second equation from the first one given in the problem. Thus, from Hess' law, $\Delta H = (-311$ kJ$) - (-136$ kJ$)$.
$\Delta H = -175$ kJ

CHAPTER TWELVE

12.45 (1) In order to calculate the heat capacity of the calorimeter, we need to know the temperature rise of the calorimeter for a particular amount of heat released. Then we can use the relationship $q = -C_{cal}\Delta T$.

The heat released in the reaction can be calculated from the internal energy change of combustion of naphthalene, ΔE_c. ΔE_c can be calculated from ΔH_c (in Table 12.3).

$q = q_v = \Delta E_c = \Delta H_c - \Delta(PV) = \Delta H_c - \Delta n_{gas}(RT)$.

In order to determine Δn_{gas}, write the balanced equation for the combustion of naphthalene:

$C_{10}H_8(s) + 12O_2(g) \to 10CO_2(g) + 4H_2O(l)$.

$\Delta n_{gas} = 10 - 12 = -2$

$\Delta E_c = \Delta H_c - \Delta n_{gas}(RT)$

$= (-5153.9 \text{ kJ/mol}) - (-2)(8.314 \times 10^{-3} \text{ kJ/mol·K})(298K)$

$\Delta E_c = -5148.9 \text{ kJ/mol}$

(2) In order to calculate the actual heat released, we calculate the moles of naphthalene combusted:

$(0.605 \text{ g})(1 \text{ mol}/128.17 \text{ g}) = 4.72 \times 10^{-3} \text{ mol}$

$\Delta E = (4.72 \times 10^{-3} \text{ mol naphth.})(-5148.9 \text{ kJ/mol naphth.}) = -24.3 \text{ kJ} = q$

$C_{cal} = -(q/\Delta T) = -(-24.3 \text{ kJ})/(2.255°C) = \mathbf{10.8 \text{ kJ/°C}}$

(3) In order to calculate the heat of combustion for the unknown compound, we need to know the heat released upon combustion of a specific mass of compound. The heat released can be calculated from the observed temperature change and the heat capacity of the calorimeter (calculated above).

$q = -C_{cal}\Delta T = -(10.8 \text{ kJ/°C})(2.030°C) = -21.9 \text{ kJ} = q_v$

$q_v = \Delta E = -(21.9 \text{ kJ})/(1.67 \text{ g}) = \mathbf{-13.1 \text{ kJ/g}}$

Since the molar mass is 176 g/mol, we have $1.67/176 = 9.49 \times 10^{-3}$ moles. Therefore the molar ΔE is

$(-21.9 \text{ kJ})/(9.49 \times 10^{-3}) = \mathbf{-2.31 \times 10^3 \text{ kJ/mol}}$

(Note that this is the internal energy change of combustion, not the enthalpy of combustion. In order to calculate the enthalpy of combustion, we would need the balanced equation for the combustion process, as we did in the first part of the problem. Based on the results from the first part of the problem, we can say that the enthalpy of combustion for the unknown compound is probably not very different from the internal energy change of combustion. If we knew the formula of the compound, we could calculate its enthalpy of combustion.)

CHAPTER TWELVE

12.47 a. We can calculate the quantity of heat released in the reaction from the equation
$q = -C_{cal}\Delta T$. $q = -(11.485 \text{ kJ/K})(4.305 \text{ K}) = -49.44 \text{ kJ}$
We can calculate the moles of benzoic acid (C_6H_5COOH) burned from its mass and formula mass.
$n = (1.870 \text{ g})(1 \text{ mol}/122.1 \text{ g}) = 0.01532 \text{ mol benzoic acid}$
Since this is a constant-volume calorimeter, $q = \Delta E = -49.44 \text{ kJ}/0.01531 \text{ mol}$.
$\Delta E = -3229 \text{ kJ/mol}$
In order to calculate ΔH, we need to write the balanced equation for the combustion of benzoic acid under standard conditions and at 298K:
$C_6H_5COOH(s) + 7.5O_2(g) \rightarrow 7CO_2(g) + 3H_2O(l)$.
$\Delta n_{gas} = 7 - 7.5 = -0.5$
$\Delta H = \Delta E + \Delta(PV) = \Delta E + \Delta n_{gas}(RT)$
$\Delta H = (-3229 \text{ kJ/mol}) + (-05)(8.314 \times 10^{-3} \text{ kJ/mol·K})(298K)$
      = **–3230. kJ/mol**

b. The value of ΔH for the combustion of benzoic acid is given as –3227 kJ/mol in Table 12.3. This value is very close to the value we calculated in part (a), differing only in the last significant figure.

c. If $H_2O(g)$ was a product, instead of $H_2O(l)$, we would have
$\Delta n_{gas} = 10 \text{ mol} - 7.5 \text{ mol} = 2.5 \text{ mol}$.
$\Delta H = \Delta E + \Delta n_{gas}(RT) = (-3229 \text{ kJ/mol}) + (-2.5)(8.314 \times 10^{-3} \text{ kJ/mol·K})(298K)$
$\Delta H = \mathbf{-3223 \text{ kJ/mol}}$

12.49 One kilogram of LPG contains 950 g C_3H_8 and 50 g CH_4. We can calculate the heat released from the partial combustion of each of these fuels as described in the preceding problem.
  $(950 \text{ g } C_3H_8)(1 \text{ mol}/44.10 \text{ g}) = 21.5 \text{ mol } C_3H_8$
  $(21.5 \text{ mol } C_3H_8)(-637.2 \text{ kJ/mol } C_3H_8) = -13,700 \text{ kJ}$
  $(50 \text{ g } CH_4)(1 \text{ mol}/16.04 \text{ g}) = 3.12 \text{ mol } CH_4$
  $(3.12 \text{ mol } CH_4)(607.1 \text{ kJ}/1 \text{ mol } CH_4) = -1890 \text{ kJ}$
Total heat –15,600 kJ

Fraction due to acetylene synthesis = $(-13,700)/(15,600) = \mathbf{0.878}$

The overall standard heat of reaction can be calculated from the total heat released and the total mass of fuel consumed:
$(-15,600 \text{ kJ})/(1.000 \text{ kg}) = \mathbf{-15,600 \text{ kJ/kg}}$

CHAPTER TWELVE

12.51 Since CO and H_2 are present in 1:2 stoichiometric amounts, both reactants will be consumed. The only product is liquid methanol, and so the final pressure will be the vapor pressure of methanol, which is less than one atm at 298K.
The ΔH can be obtained from heats of formation:
$\Delta H = \Delta H°_f(CH_3OH) = \Delta H°_f(CO) = -238 - (-110) = -128$ kJ/mol.
Now convert this to ΔE:
$\Delta E = \Delta H - \Delta n_{gas} RT$
$= -128$ kJ $- (-3$ mol$)(8.314 \times 10^{-3}$ kJ/mol·K$)(298$K$)$
$= -120.$ kJ/mol.

Next, determine the number of moles of reaction. Since the two gases were initially present in a 1:2 ratio, the initial pressure of CO was 22.0/3 = 7.33 atm
$n_{co} = n_{methanol} = PV/RT = (7.33)(10.0)/(0.08206)(298) = 3.00$. Therefore,
$\Delta E = 3(-120.$ kJ/mol$) = $ **$-360.$ kJ**
Since the sign is negative, the heat is transferred **out of** the reaction.

12.53 We can calculate $\Delta H°$ for this reaction from standard enthalpies of formation listed in Table 12.4: $\Delta H° = \Delta H_f°(C_2H_6(g)) - \Delta H_f°(C_2H_4(g))$.
$\Delta H° = (-84.5$ kJ/mol$) - (52.3$ kJ/mol$) = $ **-136.8 kJ/mol**

We can calculate this same value from the heats of combustion listed in Table 12.3: $\Delta H° = -\Delta H_c°(C_2H_6(g)) + \Delta H_c°(H_2(g)) + \Delta H_c°(C_2H_4(g))$. Before doing so, though, we need to realize that $\Delta H_c°(H_2(g)) = \Delta H_f°(H_2O(l))$ since both are equal to the standard enthalpy change for the reaction:
$H_2(g) + 1/2 O_2(g) \rightarrow H_2O(l)$.
Thus $\Delta H° = -\Delta H_c°(C_2H_6(g)) + \Delta H_f°(H_2O(l)) + \Delta H_c°(C_2H_4(g))$
$\Delta H° = -(-1560$ kJ/mol$) + (-286$ kJ/mol$) + (-1411$ kJ/mol$)$
$= $ **-137 kJ/mol**

12.55 For the reaction, $2WC(s) + 5O_2(g) \rightarrow 2WO_3(s) + 2CO_2(g)$,
$\Delta H° = -2392$ kJ/mol $= 2 \Delta H_f°(CO_2(g)) + 2 \Delta H_f°(WO_3(s)) - 2 \Delta H_f°(WC(s))$.
The second reaction given tells us that $\Delta H_f°(WO_3(s)) = -1675$ kJ/mol. We can find $\Delta H_f°(CO_2(g))$ from Table 12.4.
-2392 kJ/mol $= 2(-393$ kJ/mol$) + 2(-1675$ kJ/mol$) - 2 \Delta H_f°(WC(s))$
$\Delta H_f°(WC(s)) = -872$ kJ/mol

CHAPTER TWELVE

12.57 Since the experiment was done in a bomb calorimeter, $q = q_v = \Delta E = -42.0$ kJ.
First, we need to express ΔE on a "per-mole" basis; then we can calculate ΔH from the relationship
$\Delta H = \Delta E + \Delta(PV) = \Delta E + \Delta n_{gas}(RT)$.
Arachidonic acid has a formula mass of 304.47 g/mol.
$(1.00 \text{ g})(1 \text{ mol}/304.47 \text{ g}) = 3.28 \times 10^{-3}$ mol arachidonic acid.
ΔE $(-42.0 \text{ kJ})/(3.28 \times 10^{-3} \text{ mol}) = -12,800$ kJ/mol
In order to calculate ΔH, we need the balanced combustion reaction:
$C_{20}H_{22}O_2(s) + (49/2)O_2(g) \rightarrow 20CO_2(g) + 11H_2O(l)$
from which we see that $\Delta n_{gas} = (20 - 24.5) = -4.5$.
$\Delta H = \Delta E + \Delta n_{gas}(RT) = -12,800$ kJ/mol $+ (-4.5)(8.314 \times 10^{-3}$ kJ/mol·K$)(310.$K$)$
$\Delta H = -12,800$ kJ/mol (to the appropriate number of significant figures)
Finally, convert to a "per gram" basis:
$(-12,800$ kJ/mol acid$)(1$ mol acid/304.47 g acid$) = $ **42.0 kJ/g**
In other words, ΔH and ΔE are not significantly different.

12.59 a. We need to write the balanced equation for the combustion of liquid propane:
$C_3H_8(l) + 5O_2(g) \rightarrow 3CO_2(g) + 4H_2O(l)$
$\Delta H° = \Delta H_c° = 4 \Delta H_f°(H_2O(l)) + 3 \Delta H_f°(CO_2(g)) - \Delta H_f°(C_3H_8(l))$
$\Delta H_c° = 4(-286$ kJ/mol$) + 3(-393$ kJ/mol$) - (-119$ kJ/mol$)$
  $= $ **−2204 kJ/mol**

b. The standard heat of formation of gaseous propane from Table 12.4 is −2220 kJ/mol. (The small difference between the heats of combustion for gaseous and liquid propane is due to the heat of vaporization for propane. It is approximately 15 kJ/mol.)
Since one cubic foot of propane contains approximately one mole of propane, the heat of combustion per cubic foot is the same as the heat of combustion per mole. $\Delta H_c° = $ **−2220 kJ/ft³**

c. Let us assume that the water must be heated by about 50K. Then we can calculate the quantity of heat required and the volume of propane needed to supply that heat.
In order to use the equation $q = mC\Delta T$, we need to know the mass of 100 gallons of water: $(100$ gal$)(4$ qt/1 gal$)(0.946$ L/1 qt$)(1000$ mL/1 L$)(1$ kg/L$) = 378$ kg
$q = (378$ kg$)(4.18$ kJ/kg·K$)(50$ K$) = 8 \times 10^4$ kJ of heat required.
In order to produce this quantity of heat, $(8 \times 10^4$ kJ$)(1$ft³$/2220$ kJ$) = $ **40 ft³** of propane gas must be combusted.

CHAPTER TWELVE

12.61 Let us assume that a one-liter sample of air has been collected for analysis. This means that there will be 10^{-6} L of CO present for every 1 ppm of CO. Assuming "normal" conditions of 298K and 1 atm, we can calculate the number of moles of CO present for every ppm of CO.

n = PV/RT = (1 atm)(10^{-6} L)/(0.08206 L·atm/mol·K)(298·K) = 4×10^{-8} mol CO.

In order to calculate the temperature rise upon oxidation of the CO, we need to know the heat of combustion of CO:

$CO(g) + (1/2)O_2(g) \rightarrow CO_2(g)$.

$\Delta H° = \Delta H_c° = \Delta H_f°(CO_2(g)) - \Delta H_f°(CO(g)) = (-393$ kJ/mol$) - (-110$ kJ/mol$)$

$\Delta H_c° = -283$ kJ/mol

The actual heat released upon oxidation of 4×10^{-8} mol CO can be calculated as follows:

$(-283$ kJ/mol$)(4 \times 10^{-8}) = -1.13 \times 10^{-5}$ kJ = -0.0113 J

Finally, the temperature rise can be calculated from the known heat capacity for air: $q = nC\Delta T$. In one liter of air at 298K and 1 atm, there are 0.0409 moles of air present:

n = (1 atm)(1 L)/(0.08206 L·atm/mol·K)(298K) = 0.0409 mol air.

q = 0.0113 J = (0.0409 mol)(20.8 J/mol·K)(ΔT); **$\Delta T = 1.33 \times 10^{-2}$ K**

The temperature of the air will rise by approximately 1.33 hundredths of a degree for every ppm of CO. This will require a very sensitive thermometer, indeed!

12.69 a. First calculate the energy released:

$(3.2 \times 10^{-11}$ J/atom$)(6.023 \times 10^{23}$ atoms/mol$)(0.02)(100$ moles$) =$

3.8×10^{13} J energy released. Of this, 10% is converted to kinetic energy:

KE = 3/2 nRT = 3.8×10^{12} J.

Now solve for the temperature:

T = 2E/3nR = $(2)(3.8 \times 10^{12}$ J$)/[3(25000$ mol$)(8.314$ J/mol·K$)]$

= **1.2×10^7 K**

b. The rms velocity is

$v = \sqrt{3RT/M}$ where all quantities expressed in SI units will give v in m/s.

$v = \sqrt{(3)(8.314)(1.2 \times 10^7)/0.235}$

= 3.6×10^4 m/s = **3.6×10^6 cm/s**

c. P = nRT/V = (25,000 mol)(0.08206 L·atm/mol·K)(12×10^7 K)/10.000 L

= **2.5×10^6 atm**

Note: Although results are carried out to 2 significant figures, the problem really warrants only one.

CHAPTER THIRTEEN

Spontaneous Change

13.1 Work of expansion against a constant pressure is $w = -P_{ex}\Delta V$.
$w = -(1.5 \text{ atm})(3.0 \text{ L} - 1.0 \text{ L}) = -(1.5 \text{ L·atm})(101.3 \text{ J/L·atm})$
$= \mathbf{-1.5 \times 10^2 \text{ J}}$.

w is maximum for a reversible expansion, so the work just calculated will be **less** than that for a reversible expansion.

13.3 For each step of each process, we can calculate the work from the equation $w = -P_{ex}\Delta V$, where P_{ex} = the constant external pressure against which the gas expands (the final pressure of the gas expansion).
For the one-step process: $P_1 = 1.00$ atm, $V_1 = 2.00$ L;
$V_2 = 3.00$ L, $P_2 = P_1V_1/V_2 = 0.667$ atm
$w = -P_{ex}\Delta V = -(0.667 \text{ atm})(1.00 \text{ L}) = -0.667 \text{ L·atm} = \mathbf{-67.6 \text{ J}}$
For the two-step process: $P_1 = 1.00$ atm, $V_1 = 2.00$ L;
$V_2 = 2.50$ L, $P_2 = P_1V_1/V_2 = 0.800$ atm; $V_3 = 3.00$ L, $P_3 = 0.667$ atm
$w = -(0.800 \text{ atm})(0.50 \text{ L}) - (0.667 \text{ atm})(0.50 \text{ L}) = -0.734 \text{ L·atm} = \mathbf{-74.3 \text{ J}}$
For the three-step process: $P_1 = 1.00$ atm, $V_1 = 2.00$ L, $V_2 = 2.33$ L, $P_2 = 0.858$ atm; $V_3 = 2.67$ L, $P_3 = 0.749$ atm; $V_4 = 3.00$ L, $P_4 = 0.667$ atm.
$w = -(0.858 \text{ atm})(0.33 \text{ L}) - (0.749 \text{ atm})(0.34 \text{ L}) - (0.667 \text{ atm})(0.33 \text{ L})$
$w = -0.758 \text{ L·atm} = \mathbf{-76.8 \text{ J}}$
The work obtained from the system (−w) increases with the number of irreversible steps taken. Increasing the number of steps taken approaches reversibility so these calculations are consistent with the statement that $-w_{rev}$ is the greatest quantity of work obtainable for the change between initial and final states.

13.5 The melting of ice at its normal temperature is reversible,
$\Delta S_{fus} = \Delta H_{fus}/T_{mp} = (6025 \text{ J/mol})/273 \text{K}$
$= \mathbf{22.1 \text{ J/mol·K}}$
$\Delta S_{univ} = \mathbf{0}$ for a reversible process

13.7 The melting of lead at 327°C is reversible so $\Delta S_{univ} = \mathbf{0}$, and $q = q_{rev}$.
$(1.00 \times 10^2 \text{ g Pb})(1 \text{ mol}/207.2 \text{ g}) = 4.83$ mol Pb
Melting is endothermic so $q_{rev} = (21.3 \text{ kJ/mol})(4.83 \text{ mol}) = 103$ kJ.
$\Delta S = q_{rev}/T = (103 \times 10^3 \text{ J})/(600. \text{K}) = \mathbf{172 \text{ J/K}}$

CHAPTER THIRTEEN

13.9 We can obtain $\Delta G°$ from $\Delta G° = \Delta H° - T\Delta S°$. Be sure units are consistent.

$\Delta G° = -1220$ kJ/mol $- (298K)(-0.130$ kJ/mol·K$)$
$= -1220 + 38.7 = $ **-1181 kJ**

Since $\Delta G°$ is negative, the process is spontaneous under standard conditions.

13.11 a. $\Delta S° = [2(40) - (205) - 2(41.6)]$ J/mol·K = **-208.2 J/mol·K**
$\Delta G° = [2(-604.2) - (0) - 2(0)]$ kJ/mol = **-1208 kJ/mol**
$\Delta H° = (-1208$ kJ/mol$) + (298K)(-0.2082$ kJ/mol·K$) = $ **-1270 kJ/mol**

b. $\Delta S° = [(270) - 4(130.6) - 3(5.73)]$ J/mol·K = **-269.6 J/mol·K**
$\Delta G° = [(-23.5) - 4(0) - 3(0)]$ kJ/mol = **-23.5 kJ/mol**
$\Delta H° = (-23.5$ kJ/mol$) + (298K)(-0.2696$ kJ/mol·K$) = $ **-103.8 kJ/mol**

c. $\Delta S° = [(165.1) + (114.6) - (187)]$ J/mol·K = **92.7 J/mol·K**
$\Delta G° = [(203.2) + (105.4) - (-95.27)]$ kJ/mol = **403.9 kJ/mol**
$\Delta H° = (403.9$ kJ/mol$) + (298K)(0.0927$ kJ/mol·K$) = $ **431.5 kJ/mol**

d. $\Delta S° = [(127) - 1/2(205) - (186)]$ J/mol·K = **-161.5 J/mol·K**
$\Delta G° = [(-166.2) - 1/2(0) - (-50.79)]$ kJ/mol = **-115.4 kJ/mol**
$\Delta H° = (-115.4$ kJ/mol$) + (298K)(-0.1615$ kJ/mol·K$) = $ **-163.5 kJ/mol**

e. $\Delta S° = [(198) + (189) - (214) - (130.6)]$ J/mol·K = **42.4 J/mol·K**
$\Delta G° = [(-137.3) + (-228.6) - (-394.4) - (0)]$ kJ/mol = **28.5 kJ/mol**
$\Delta H° = (28.5$ kJ/mol$) + (298K)(0.0424$ kJ/mol·K$) = $ **41.1 kJ/mol**

a. $\Delta S° = [3(31.9) + 2(189) - (248) - 2(205)]$ J/mol·K = **-184.3 J/mol·K**
$\Delta G° = [3(0) + 2(-228.6) - (-300.4) - 2(-33.0)]$ kJ/mol = **-90.8 kJ/mol**
$\Delta H° = (-90.8$ kJ/mol$) + (298K)(-0.1843$ kJ/mol·K$) = $ **-145.7 kJ/mol**

13.13 a. $\Delta G° = $ **-1208 kJ/mol**

Since $\Delta G° < 0$, the reaction is spontaneous as written.

b. $\Delta G° = $ **-23.5 kJ/mol**

Since $\Delta G° < 0$, the reaction is spontaneous as written.

c. $\Delta G° = $ **403.9 kJ/mol**

Since $\Delta G° > 0$, the reaction is not spontaneous as written.

d. $\Delta G° = $ **-115.4 kJ/mol**

Since $\Delta G° < 0$, the reaction is spontaneous as written.

e. $\Delta G° = $ **28.5 kJ/mol**

Since $\Delta G° > 0$, the reaction is not spontaneous as written.

f. $\Delta G° = $ **-90.8 kJ/mol**

Since $\Delta G° < 0$, the reaction is spontaneous as written.

CHAPTER THIRTEEN

13.15 a. $\Delta G° = [4(-127) - (0) - 2(-146)]$ kJ/mol = **–216 kJ/mol**
b. $\Delta G° = [(-228.6) + (68.12) - (-174.8)]$ kJ/mol = **14.3 kJ/mol**
c. $\Delta G° = [(-23.5) - (0) - (62.84)]$ kJ/mol = **–86.3 kJ/mol**

13.17 a. $K = e^{-(-216,000 \text{ J/mol})/(8.314 \text{ J/mol·K})(298K)} = \mathbf{7.29 \times 10^{37}}$
b. $K = e^{-(14,300 \text{ J/mol})/(8.314 \text{ J/mol·K})(298K)} = \mathbf{3.11 \times 10^{-3}}$
c. $K = e^{-(-86,300 \text{ J/mol})/(8.314 \text{ J/mol·K})(298K)} = \mathbf{1.34 \times 10^{15}}$

13.19 In order to calculate $\Delta G°$ at 683K, we need to know $\Delta H°$ and $\Delta S°$ for this reaction. First, we will calculate $\Delta S°$ using Table 13.1. Then we will calculate $\Delta G°(298)$ using Table 13.2. Finally, we will calculate $\Delta H°$ using $\Delta G°(298) = \Delta H° - (298K) \Delta S°$.

$\Delta S° = (205 \text{ J/mol·K}) + 2(191 \text{ J/mol·K}) - 2(220 \text{ J/mol·K}) = 147$ J/mol·K
$\Delta G° = (298) = (0 \text{ kJ/mol}) + 2(0 \text{ kJ mol}) - 2(104 \text{ kJ/mol}) = -208$ kJ/mol
$\Delta H° = (-208 \text{ kJ/mol}) + (298K)(0.147 \text{ kJ/mol·K}) = -164$ kJ/mol

Now we can use $\Delta G°(683) = \Delta H° - (683K) \Delta S°$. We are assuming that $\Delta H°$ and $\Delta S°$ remain constant with temperature.
$\Delta G°(683) = (-164 \text{ kJ/mol}) - (683K)(0.147 \text{ kJ/mol·K}) = \mathbf{-264}$ **kJ/mol**
Now that we know $\Delta G°$ at 683K, we can use it to calculate the equilibrium constant at 683K:
$K = e^{-\Delta G°/RT} = e^{-(-264,000 \text{ J/mol})/(8.314 \text{ J/mol·K})(683K)}$
$\mathbf{K = 1.55 \times 10^{20}}$

13.21 $\Delta G° = \Delta H° - T\Delta S°$. We will take $\Delta H°$ and $\Delta S°$ to be independent of temperature.
$\Delta G° = 1170$ kJ/mol $- (150.K)(0.445$ kJ/mol·K$)$
$= 1170 - 66.8 = \mathbf{1103}$ **kJ**
Reaction is spontaneous in the **reverse direction**.

13.23 All participating species in their standard states means that all gases will be at 1 atm pressure at equilibrium. This means that all the equilibrium constants will be equal to one. The question, then, is to find the temperature at which each reaction has K = 1.
When K = 1, $\Delta G° = 0$, $\Delta H° = T\Delta S°$ and $T = \Delta H°/\Delta S°$. We will use Tables 13.2 and 13.1 to solve for $\Delta G°$ and $\Delta S°$ and then solve for the temperature of interest.
a. $\Delta G° = (-23.5$ kJ/mol$) - (62.34$ kJ/mol$) = -85.8$ kJ/mol
$\Delta S° = (270$ J/mol·K$) + (130.6$ J/mol·K$) - (266.9$ J/mol·K$) = -128$ J/mol·K
$\Delta H° = \Delta G° + (298K)\Delta S° = -124$ kJ/mol
$T = \Delta H°/\Delta S° = (-124000$ J/mol$)/(-128$ J/K mol$) = \mathbf{969K}$

CHAPTER THIRTEEN

b. $\Delta G° = (-394.4 \text{ kJ/mol}) + (-50.79 \text{ kJ/mol}) - (-392 \text{ kJ/mol}) = -53 \text{ kJ/mol}$
$\Delta S° = (214 \text{ J/mol·K}) + (186 \text{ J/mol·K}) - (160 \text{ J/mol·K}) = 240 \text{ J/mol·K}$
$\Delta H° = \Delta G° + (298K)\Delta S° = 19 \text{ kJ/mol}$
$T = \Delta H°/\Delta S° = (19000 \text{ J/mol})/(240 \text{ J/K·mol}) = \mathbf{79K}$

13.25 In order to solve this problem, we will make use of the relationship
$\ln(K_2/K_1) = -(\Delta H°/R)(1/T_2 - 1/T_1)$.
$\Delta H° = -R\ln(K_2/K_1)/(1/T_2 - 1/T_1) = R[\ln(K_1/K_2)][T_1T_2/(T_1-T_2)]$
$\Delta H° = (8.314 \text{ J/mol·K})[\ln(3.05 \times 10^3/4.52 \times 10^2)][(450K)(298K)/(152K)]$
$\Delta H° = 8.16 \times 10^4 \text{ J/mol} = \mathbf{81.6 \text{ kJ/mol}}$

13.27 In order to determine the best conditions for performing these reactions, we need to determine the sign of $\Delta H°$ and the overall production or consumption of gas molecules.
$CaCO_3(s) \rightarrow CaO(s) + CO_2(g)$
$\Delta S° = (214 \text{ J/mol·K}) + (40 \text{ J/mol}) - (92.9 \text{ J/mol·K}) = 161 \text{ J/mol·K}$
$\Delta G° = (-394.4 \text{ kJ/mol}) + (-604.2 \text{ kJ/mol}) - (-1129 \text{ kJ/mol}) = 130.4 \text{ kJ/mol}$
$\Delta H° = \Delta G° + (298K)\Delta S° = 178.4 \text{ kJ/mol}$.

The forward reaction is endothermic and will be favored by **high temperatures**. It also produces a gas (CO_2) and so will be favored by **low pressures**.
$CaO(s) \rightarrow H_2O(liq) + Ca(OH)_2(s)$
$\Delta S° = (72.8 \text{ J/mol·K}) - (70.00 \text{ J/mol·K}) - (40 \text{ J/mol·K}) = -37.2 \text{ J/mol·K}$
$\Delta G° (-896.6 \text{ kJ/mol}) - (-237.2 \text{ kJ/mol}) - (-604.2 \text{ kJ/mol}) = -55.2 \text{ kJ/mol}$
$\Delta H° = \Delta G° + (298K)\Delta S° = -66.3 \text{ kJ/mol}$.

This reaction is exothermic and will be favored by **low temperatures**. There are no gases produced or consumed so it will **not be affected much by pressure**.

13.29 The Haber Process for the preparation of ammonia from its elements is
$N_2(g) + 3H_2(g) \rightarrow 2NH_3(g)$
The heat of formation of ammonia is -46.0 kJ/mol (Table 12.4a). Thus, the forward reaction is exothermic, and it consumes more molecules of gas than it produces (see Section 10.4 in the text). The Haber Process will yield the most product under conditions of **high pressure and low temperature**.

CHAPTER THIRTEEN

13.31 According to the Clausius-Clapeyron equation, $\ln(p_2/p_1) = -(\Delta H_{vap}/R)[(1/T_2) - (1/T_1)]$, where p_2 and p_1 are the vapor pressures of a liquid at T_2 and T_1, respectively, ΔH_{vap} is the enthalpy of vaporization for the liquid and R is the ideal gas constant.
For hexane, we will let p_1 = 100.0 torr, p_2 = 40.0 torr, T_1 = 289.0K and T_2 = 270.9K.
$\Delta H_{vap} = -R\ln(p_2/p_1)/(1/T_2 - 1/T_1) = R[\ln(p_1/p_2)][T_1T_2/(T_1 - T_2)]$
ΔH_{vap} = (8.314 J/mol·K) [ln(100.0/40.0)][(289.0)(270.9)/18.1]
$\Delta H_{vap} = 3.30 \times 10^4$ J/mol = **33.0 kJ/mol**

13.33 a. False: G = H − TS b. True c. False
d. False: H = E + PV e. True f. True

13.35 ΔH_{vap} for water can be calculated from the heats of formation in Table 12.4 For the process H_2O(liq) → H_2O(g), ΔH = (−242 kJ/mol) − (−286 kJ/mol) = 44 kJ/mol = ΔH_{vap}.
Since this is a constant-pressure process, **ΔH = q = 44 kJ.**
$\Delta S_{vap} = \Delta H_{vap}/T_b$ = (44,000 J/mol)/(373K) = **120 J/mol·K**.
This is a reversible process, so **ΔG = 0**.
Since this is a constant-pressure process,
w = $-P_{ex}\Delta V = -P_{ex}(V_{gas} - V_{liq})$. The volume of one mole of liquid water can be calculated from the known density of water:
V_{liq} = (1 mol)(18.02 g/mol)(1 mL/1 g) = 18 mL. The volume of one mole of gaseous water can be calculated from the ideal gas equation:
V_{gas} = nRT/P = (1 mol)(0.08206 L·atm/mol·K)(373K)/(1.00 atm) = 30.6 L.
So $(V_{gas} - V_{liq}) = V_{gas}$ = 30.6 L.
w = −(1.00 atm)(30.6 L) = −30.6 L·atm = **−3.1 kJ**
Since we have already calculated the work and heat, we can calculate ΔE from ΔE = q + w = 44 kJ + (−3.1 kJ) = **41 kJ = ΔE**.

13.37 a. $\Delta G° = \Delta H° - T\Delta S°$
∴ $\Delta S° = (\Delta H° - \Delta G°)T$ = (−146 kJ/mol − (−8.4 kJ/mol)/298K
$\Delta S°_f$ = **−462 J/mol·K**

b. n-pentane has an absolute entropy of 349 J/mol·K. The formation reaction is written as follows:
$6H_2(g) + 5C(s) \rightarrow C_5H_{12}(l)$. Therefore, $\Delta S = S_{C_5H_{12}} - 6S_{H_2} - 5S_C$
Neglecting the entropy of carbon, $\Delta S = -462 = 349 - 6S_{H_2}$
S_{H_2} = **135 J/mol·K**

CHAPTER THIRTEEN

13.39 $\Delta G°$ is related to K_{sp} by $\Delta G° = -RT\ln K_{sp}$
$\Delta G° = -(8.314 \text{ J/mol·K})(291\text{K})\ln(5.0 \times 10^{-5}) = +24.0 \text{ kJ/mol}$
$\Delta G° = \Delta H° - T\Delta S°$; $\Delta S° = (\Delta H° - \Delta G°)/T = (-42.0 - 24.0)/291\text{K}$
$\Delta S° = 0.23 \text{ kJ/mol·K} = $ **230 J/mol·K**

13.41 In order to calculate the equilibrium constant at 298K for the given reaction, we need to know $\Delta G°(298)$. This can be calculated from the data in Table 13.2:
$\Delta G° = 4(51.84 \text{ kJ/mol}) - 3(0) - 2(104.18 \text{ kJ/mol}) = -1.00 \text{ kJ/mol}$
$K = e^{-\Delta G°/RT} = e^{-(-1000 \text{ J/mol})/(8.314 \text{J/mol·K})(298\text{K})} = $ **1.5**
In an equilibrium mixture containing 0.500 atm each of N_2O and NO_2, we can calculate the partial pressure of O_2 by setting up the equilibrium expression:
$1.5 = (p_{NO_2})^4/p_{O_2})^3(p_{N_2O})^2 = (0.500)^4/(p_{O_2})^3(0.500)^2$
$(p_{O_2})^3 = (0.500)^2/1.5 = 0.167$; $p_{O_2} = $ **0.55 atm**

13.43 a. $K_p = p_{SO_3}/p_{SO_2}(p_{O_2})^{1/2}$

b. In order to calculate $\Delta H°$ and $\Delta S°$ algebraically, we will use the two points at the extremes of temperature and make use of the equations
$\ln(K_2/K_1) = -(\Delta H°/R)(1/T_2 - 1/T_1)$ and $\ln K = -(\Delta H°/RT) + (\Delta S°/R)$.
We can calculate $\Delta H°$ from the first equation. For the points we have chosen,
$T_2 = 1100\text{K}$, $K_2 = 0.628$, $T_1 = 800\text{K}$ and $K_1 = 31.3$.
$\ln(0.628/31.3) = -(\Delta H°/8.314 \text{ J/mol·K}[(1/1100\text{K}) - (1/800\text{K})]$
$\Delta H° = -95,300 \text{ J/mol} = $ **−95.3 kJ/mol**
We can calculate $\Delta S°$ from the second equation. We will use $T = 800\text{K}$ and $K = 31.3$.
$\ln(31.3) = -(-95,300 \text{ J/mol})/[(8.314 \text{ J/mol·K})(800\text{K})] + \Delta S°/(8.314 \text{ J/mol·K})$
$\Delta S° = $ **−90.5 J/mol·K**

c. In order to solve for $\Delta H°$ and $\Delta S°$ graphically, we will make use of
$\ln K = -(\Delta H°/RT) + (\Delta S°/R)$. This equation is of the form $y = mx + b$ where $y = \ln K$, $x = (1/T)$, $m = -\Delta H°/R$ and $b = \Delta S°/R$. A plot of $\ln K$ vs. $1/T$, then, should yield a straight line graph with slope $= -\Delta H°/R$ and y-intercept $= \Delta S°/R$.

T(K)	1/T (K^{-1})	ln K
800	1.25×10^{-3}	3.444
850	1.18×10^{-3}	2.625
900	1.11×10^{-3}	1.879
950	1.05×10^{-3}	1.176
1000	1.00×10^{-3}	0.615
1100	9.09×10^{-3}	−0.465

CHAPTER THIRTEEN

13.45 In order to calculate $\Delta G°$ at these different temperatures, we need to know $\Delta H°$ and $\Delta S°$. Since the reaction is written as twice the formation reaction of $NH_3(g)$, these thermodynamic parameters can be calculated simply from Tables 12.4 and 13.1.
$\Delta H° = 2(-46.0 \text{ kJ/mol}) = -92.0 \text{ kJ/mol}$
$\Delta S° = 2(192 \text{ J/mol}) - 3(130.6 \text{ J/mol·K}) - (191 \text{ J/mol·K})$
$= -198.8 \text{ J/mol·K}$

At 0°C:
$\Delta G° = (-92.0 \text{ kJ/mol}) - (273K)(-0.1988 \text{ kJ/mol·K}) = -37.7 \text{ kJ/mol}$
$K = e^{-\Delta G°/RT} = e^{-(-37,700)/(8.314)(273)} = \mathbf{1.64 \times 10^7}$

At 200°C:
$\Delta G° = (-92.0 \text{ kJ/mol}) - (473K)(-0.1988 \text{ kJ/mol·K}) = 2.0 \text{ kJ/mol}$
$K = e^{-\Delta G°/RT} = e^{-(2000)/(8.314)(473)} = \mathbf{0.60}$

At 400°C:
$\Delta G° = (-92.0 \text{ kJ/mol}) - (673K)(-0.1988 \text{ kJ/mol·K}) = 41.8 \text{ kJ/mol}$
$K = e^{-\Delta G°/RT} = e^{-(41,800)/(8.314)(673)} = \mathbf{5.70 \times 10^{-4}}$

Increasing the temperature results in a decrease in the equilibrium constant for the reaction. This is consistent with our earlier assertion that for an exothermic reaction, a temperature increase will shift the equilibrium to the left. Increasing the temperature limits the extent of conversion of reactants to products.

13.47 Osmosis is the process of water passing through a membrane into a solution and making it more dilute. It is a spontaneous process. Reverse osmosis is not a spontaneous process in the thermodynamic sense. A salt solution may be forced through a semi-permeable membrane to yield pure water. Pressure is applied to accomplish this, and work is performed on the system. There is nothing in the second law that forbids doing work to bring about an otherwise non-spontaneous process.

CHAPTER FOURTEEN

Electrochemistry

14.1 a. Chlorine is +1 in Cl_2O, +4 in ClO_2 and +7 in Cl_2O_7.
  b. Sulfur is −2 in HS^-, +4 in HSO_3^-, +7 in $HS_2O_8^-$, +2.5 in $HS_4O_6^-$ and +6 in HSO_4^-
  c. Phosphorus is 0 in P_4, −3 in PH_3, −2 in P_2H_4, +1 in H_3PO_2, +3 in H_3PO_3, +5 in H_3PO_4 and +5 in $(NH_4)_4P_2O_7$.

14.3 a. Na is the reducing agent. Cl_2 is the oxidizing agent.
  b. Zn is the reducing agent. H_2SO_4 is the oxidizing agent.
  c. Fe is the reducing agent, Cl_2 is the oxidizing agent.
  d. S is the reducing agent, F_2 is the oxidizing agent.

14.5 a. Iron is **oxidized**.
  b. Sulfide is **oxidized**.
  c. Sulfite is **oxidized**.
  d. Persulfate is **reduced**.

14.7 a. $14H^+ + Cr_2O_7^{2-} + 6Cl^- \rightarrow 2Cr^{3+} + 3Cl_2 + 7H_2O$
  b. $4H^+ + MnO_2 + 2Hg + 2Cl^- \rightarrow Mn^{2+} + Hg_2Cl_2 + 2H_2O$
  c. $4H^+ + 3Ag + NO_3^- \rightarrow 3Ag^+ + NO + 2H_2O$
  d. $8H^+ + H_3AsO_4 + 4Zn \rightarrow AsH_3 + 4Zn^{2+} + 4H_2O$
  e. $18H_2O + 10Au^{3+} + 3I_2 \rightarrow 10Au + 6IO_3^- + 36H^+$
  f. $6H^+ + IO_3^- + 8I^- \rightarrow 3I_3^- + 3H_2O$
  g. $H^+ + 3HS_2O_3^- \rightarrow 4S + 2HSO_4^- + H_2O$
  h. $4H^+ + 2O_2^{2-} \rightarrow O_2 + 2H_2O$

14.9 a. $OH^- + 2Co(OH)_3 + Sn \rightarrow 2Co(OH)_2 + HSnO_2^- + H_2O$
  b. $3ClO_4^- + I^- \rightarrow 3ClO_3^- + IO_3^-$
  c. $OH^- + H_2O + PbO_2 + Cl^- \rightarrow ClO^- + Pb(OH)_3^-$
  d. $OH^- + H_2O + NO_2^- + 2Al \rightarrow NH_3 + 2AlO_2^-$
  e. $2ClO^- \rightarrow 2Cl^- + O_2$
  f. $2OH^- + HXeO_4^- + 3Pb \rightarrow Xe + 3HPbO_2^-$
  g. $2H_2O + 2Ag_2S + 8CN^- + O_2 \rightarrow 2S + 4Ag(CN)_2^- + 4OH^-$
  h. $8H_2O + 2MnO_4^- + 7S^{2-} \rightarrow 2MnS + 5S + 16OH^-$
  i. $2OH^- + 2ClO_2 \rightarrow ClO_2^- + ClO_3^- + H_2O$

CHAPTER FOURTEEN

14.11 The more negative the reduction potential, the better the reducing agent:
Na > Al > Co > Ni > H_2 > Ag

14.13 a. **I_2** could oxidize Fe → Fe^{3+}.
   b. **I_2** could oxidize Fe → Fe^{2+}.
   c. **Br_2** could oxidize Fe^{2+} to Fe^{3+}.
   d. **H_4XeO_6** could oxidize $2F^-$ → F_2.

   Note: There are many correct answers to this question. Each answer given is just one possibility.

14.15 a. E° = −0.77 V + 1.51 V = 0.74 V
   Since E° > 0, the reaction will be **spontaneous** under standard conditions.
   b. E° = −0.22 V + 1.23 V = 1.01 V
   Since E° > 0, the reaction will be **spontaneous** under standard conditions.
   c. E° = 0.138 V + 0.268 V = 0.406 V
   Since E° > 0, the reaction will be **spontaneous** under standard conditions.
   d. E° = −0.560 V + 0.558 V = 0.002 V
   Since E° > 0, the reaction will be **spontaneous** under standard conditions.

14.17 Mn + Ni^{2+} → Mn^{2+} + Ni
   E° = 1.18 V − 0.250 V = 0.93 V

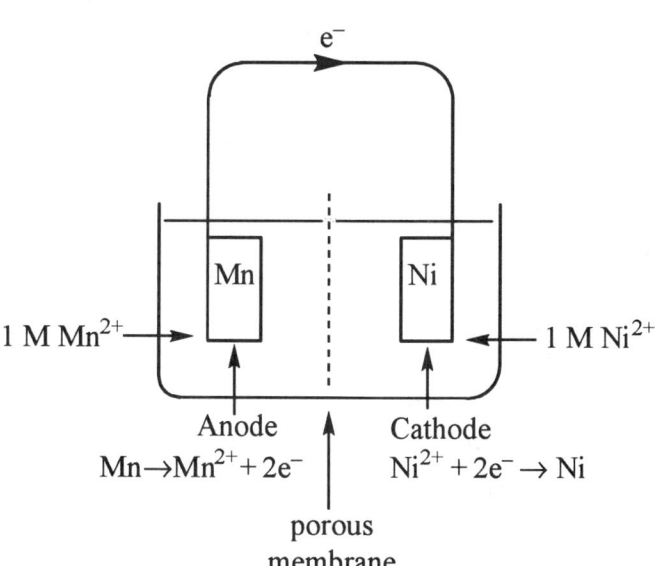

CHAPTER FOURTEEN

14.19 $2Cr + 3Cu^{2+} \rightarrow 2Cr^{3+} + 3Cu$

$E° = 0.744$ V $- 0.337$ V $= 0.407$ V

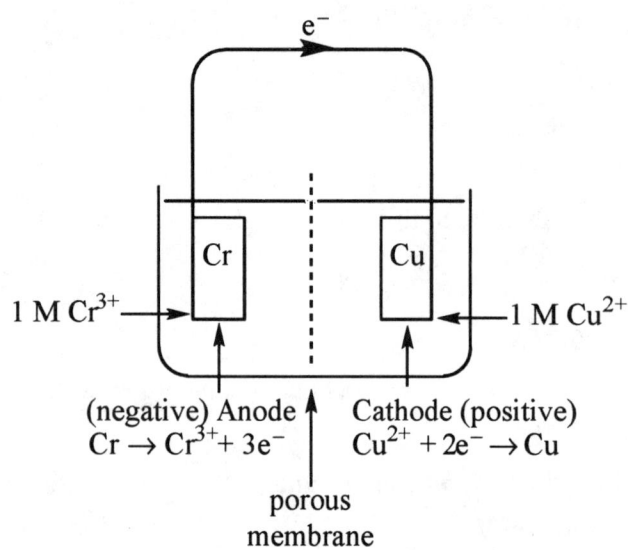

14.21 The anode reaction is:
$Zn + 2KOH + 2OH^- \rightarrow K_2ZnO_2 + 2e^- + 2H_2O$

The cathode reaction is:
$H_2O + 2e^- + HgO \rightarrow Hg + 2OH^-$

The overall cell reaction is:
$Zn + 2KOH + HgO \rightarrow K_2ZnO_2 + Hg + H_2O$

14.23 $\Delta G° = -nFE°$

$\Delta G° = -(1 \text{ mol } e^-/\text{mol of reaction})(96{,}500 \text{ C/mol } e^-)(1.27 \text{ J/C})$
$= \mathbf{-122 \text{ kJ/mol}}$

14.25 $Sn \rightarrow Sn^{2+} + 2e^-$ $E° = 0.138$ V

$\underline{Pb^{2+} + 2e^- \rightarrow Pb \quad\quad E° = -0.126 \text{ V}}$

overall: $Sn + Pb^{2+} \rightarrow Sn^{2+} + Pb$ $E° = 0.012$ V

$\Delta G° = -nFE° = -(2 \text{ mol } e^-/\text{mol reaction})(96{,}500 \text{ coul/mol } e^-)(0.012 \text{ V})$
$= -2316 \text{ J/mol} = \mathbf{-2.32 \text{ kJ/mol}}$

$K = e^{-\Delta G°/RT} = e^{-(-2316 \text{ J/mol})/(8.314 \text{ J/mol·K})(298K)}$

$\mathbf{K = 2.55}$

CHAPTER FOURTEEN

14.27 $2Ag + Br_2 \rightarrow 2Br^- + 2Ag^+$

$E° = -0.799 \text{ V} + 1.087 \text{ V} = 0.288 \text{ V}$

The maximum possible work available from such a system, $w_{elec} = -\Delta G° = nFE°$.

$w_{elec} = (2 \text{ mol e}^-/2 \text{ mol Ag})(1.00 \text{ mol Ag})(96,500 \text{ coul/mol e}^-)(0.288 \text{ V})$

$w_{elec} = 27792 \text{ J} = \textbf{27.8 kJ}$

14.29 a. $2Fe^{3+} + Zn \rightarrow 2Fe^{2+} + Zn^{2+}$

b. The Pt electrode is the cathode and the Zn electrode is the anode. Since this is a galvanic cell, the Pt electrode is positive and the Zn electrode is negative.

c. $E° = 0.77 \text{ V} - (-0.763 \text{ V}) = \textbf{1.53 V}$

$K = 10^{(2E°/0.0591)} = \textbf{10}^{\textbf{51.9}}$

d. If [Fe^{3+}] were to increase, Q would decrease and E would become more positive.

14.31 $Zn^{2+} + 2e^- \rightarrow Zn(s)$ $E° = -0.763 \text{ V}$

$Cr^{3+} + 3e^- \rightarrow Cr(s)$ $E° = -0.744 \text{ V}$

The spontaneous (galvanic) reaction is.
$3Zn(s) + 2Cr^{3+} \rightarrow 3Zn^{2+} + 2Cr(s)$

$E° = -0.744 \text{ V} - (-0.763 \text{ V}) = 0.019 \text{ V}$

Next, use the Nernst equation. Since we are looking for the equilibrium concentrations, E = 0. Let x = Cr^{3+}. Then

$E = E° - 0.0591/n \log [Zn^{2+}]^3[Cr^{3+}]^2$

$0 = 0.019 - (0.0591/6)\log [0.132]^3/[x]^2$

$-1.93 = \log [0.132]^3/[x]^2$

$10^{-1.93} = [0.132]^3/[x]^2 = 1.17 \times 10^{-2} = (2.30 \times 10^{-3})/[x]^2$

$x = \sqrt{2.30 \times 10^{-3}/1.17 \times 10^{-2}} = [Cr^{3+}] = \textbf{0.444 M}$

14.33 $E° = -0.744 - (-1.18) = 0.44 \text{ V}$

Next, apply the Nernst equation with n = 6.

$E = E° - (0.0591/n) \log[Mn^{2+}]/[Cr^{3+}]^2$

$E = 0.44 - (0.0591/6) \log ([0.10]^3/[0.01]^2)$

$E = \textbf{0.43 V}$

14.35 $Pb + 2Hg^{2+} \rightarrow Pb^{2+} + Hg_2^{2+}$

$E° = 0.126 \text{ V} + 0.920 \text{ V} = 1.046 \text{ V}$

$E = E° - (0.0591/2) \log Q$

$E = 1.046 \text{ V} - (0.0591/2) \log[(0.0936)(0.00235)/(0.00235)^2] = \textbf{0.999 V}$

The Pb electrode is the anode and the Pt electrode is the cathode. Since this is a galvanic cell, the Pb electrode is negative and the **Pt electrode is positive.**

CHAPTER FOURTEEN

14.37 Since the hydrogen gas electrode is positive, it must be the cathode. The two half reactions and the cell reaction are:

$$2H^+ + 2e^- \rightarrow H_2$$
$$\underline{Pb \rightarrow Pb^{2+} + 2e^-}$$

Net: $Pb + 2H^+ \rightarrow Pb^{2+} + H_2$

The Nernst equation is $E = E^o - (0.0591/2) \log Q$

$E^o = 0 - (-0.126) = 0.126$ V

$Q = [Pb^{2+}]/p_{H_2}/[H^+]^2 = [Pb^{2+}]$ since $p_{H_2} = 1$ atom and $[H^+] = 1$ M

Then $0.562 = 0.126 - 0.0296 \log [Pb^{2+}]$

Log $[Pb^{2+}] = -0.436/0.0296 = -14.73$

$[Pb^{2+}] = 1.9 \times 10^{-15}$ M

The solubility reaction for PbF_2 is:

$PbF_2 \leftrightarrow Pb^{2+} + 2F^-$

$K_{sp} = [Pb^{2+}][F^-]^2 = (1.9 \times 10^{-15})(1 \text{ M})^2$

$K_{sp} = \mathbf{1.9 \times 10^{-15}}$

14.39 In the 0.432 M cell, $Pb^{2+} + 2e^- \rightarrow Pb$.

In the 0.000149 M cell, $Pb \rightarrow Pb^{2+} + 2e^-$.

Since this is a concentration cell, $E^o = 0$. When the cell is first connected,

$E = 0 - (0.0591/2) \log (0.000149/0.432) = \mathbf{0.102 \text{ V}}$.

As the current is allowed to flow, the concentrations will start to equalize and the **cell potential will drop**. Eventually, the cell potential will drop to zero. When the cell potential reaches zero, the final concentrations of Pb^{2+} in the two cells will be equal: **0.216 M** each, assuming equal volumes of solution in the two cells.

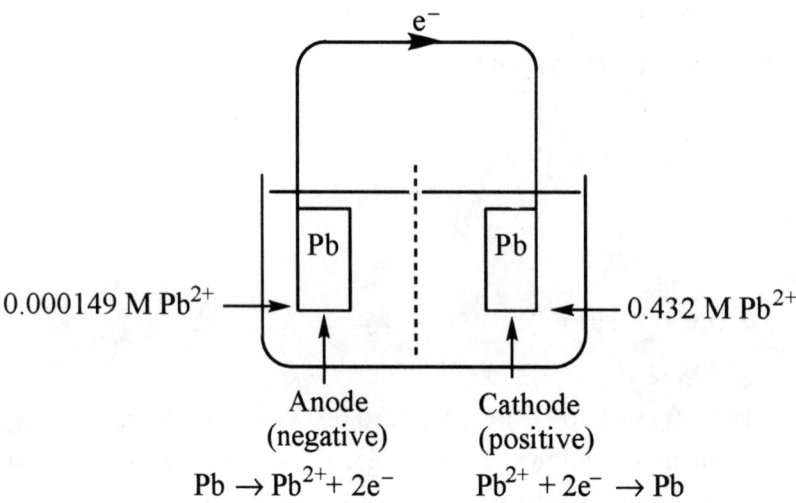

CHAPTER FOURTEEN

14.41 Since the 1 M H^+ solution contains the positive H_2 electrode, that is the solution where H^+ is being converted to H_2. This means that the unknown solution has a $[H^+] < 1$ M. This is a concentration cell so $E° = 0$.

$E = -(0.0591/n) \log [H^+]/1$

$E = 0.0251$ V $= (-0.0591/2) \log [H^+]$

$[H^+] = 0.14$ M and **pH = 0.85**

14.43 Current is charge per second: $i = q/t$
10,500 C/3600 s = **2.92 A**

14.45 The cathode half-reaction is $Cr^{3+} + 3e^- \rightarrow Cr$. We can calculate the mass of chromium that is plated from the number of electrons that flow into the cathode.
$n_{Cr} = (150{,}000 \text{ coul})(1 \text{ mol } e^-/96{,}500 \text{ coul})(1 \text{ mol Cr}/3 \text{ mol } e^-) = 0.52$ mol Cr
$(0.52 \text{ mol Cr})(51.996 \text{ g/mol}) = $ **27 g chromium**

14.47 The cathode reaction is $Co^{2+} + 2e^- \rightarrow Co$. We can calculate the current required from the mass of cobalt to be plated.
$(10.0 \text{ g Co})(1 \text{ mol}/58.9332 \text{ g}) = 0.170$ mol Co
$i = q/t = (0.170 \text{ mol Co})(2 \text{ mol } e^-/\text{mol Co})(96{,}500 \text{ coul/mol } e^-)/(86{,}400 \text{ s})$
$i = 0.380$ C/s = **0.380 A**

14.49 Electrical work is equal to the product of charge and the potential difference through which the charge must flow. We can calculate the energy required ($-w_{elec}$) from the cell potential and the total moles of electrons required.

$2Cl^- \rightarrow Cl_2 + 2e^-$, $E° = -1.36$ V

$(1000 \text{ g Cl}_2)(1 \text{ mol}/70.906 \text{ g}) = 14.1$ mol Cl_2

Energy required $= -w_{elec}$

$-w_{elec} = (14.1 \text{ mol Cl}_2)(2 \text{ mol } e^-/\text{mol Cl}_2)(96{,}500 \text{ coul/mol } e^-)(-1.36 \text{ V})$

$-w_{elec} = $ **3700 kJ**

14.51 Vanadium is in the +4 oxidation state in $VO(NO_3)_2$. $V^{4+} + 4e^- \rightarrow V$
  a. $(1.00 \times 10^4 \text{ coul})(1 \text{ mol } e^-/96{,}500 \text{ coul})(1 \text{ mol V}/4 \text{ mol } e^-) = 0.0259$ mol V
  $(0.0259 \text{ mol })(50.9415 \text{ g/mol}) = $ **1.32 g**
  b. $(1.00 \text{ coul/s})(3600 \text{ s})(1 \text{ mol } e^-/96{,}500 \text{ coul})(1 \text{ mol V}/4 \text{ mol } e^-) = 0.00933$ mol V
  $(0.00933 \text{ mol})(50.9415 \text{ g/mol}) = $ **0.475 g**
  c. 1.00 Faraday = 1.00 mol e^-; $(1.00 \text{ mol } e^-)(1 \text{ mol V}/4 \text{ mol } e^-) = 0.250$ mol V
  $(0.250 \text{ mol})(50.9415 \text{ g/mol}) = $ **12.7 g**

CHAPTER FOURTEEN

14.53 a. (2.50 coul/s)(3600 s)(1 mol/96,500 coul) = 0.0933 mol electrons
(0.0933 mol e⁻)(1 mol O_2/mol e⁻) = 0.0233 mol O_2
(0.0233 mol O_2)(22.4 L/mol) = **0.522 L O_2** collected at STP

b. The cathode reaction is: $Ag^+ + e^- \rightarrow Ag(s)$.
(0.523 g Ag)(1 mol Ag/107.9 g)(1 mol e⁻/mol Ag) = 0.00485 mol e⁻
(0.00485 mol e⁻)(96,500 coul/mol e⁻)(1 s/2.50 C) = 187 s = **3.1 min.**

14.55 n_{H_2} = PV/RT = (1 atm)(3.62 L)/(0.08206 L·atm/mol·K)(308K) = 0.143 mol H_2
$2H_2O + 2e^- \rightarrow H_2 + 2OH^-$
q = nF = (2 mol e⁻/mol H_2)(0.143 mol H_2)(96,500 C/mol e⁻)
q = 27,600 C

14.57 (1.000 g Ag)(1 mol/107.9 g) = 0.009268 mol Ag deposited
This means that 0.009268 mol e⁻ flowed.
0.009268 mol e⁻ corresponds to 0.004634 mol Cu dissolved.
(0.0043644 mol Cu)(63.546 g/1 mol) = **0.2945 g Cu** dissolved.
0.009268 mol e⁻ corresponds to 0.003089 mol Cr deposited.
(0.003089 mol)(51.996 g/1 mol) = **0.1606 g Cr** deposited.
0.009268 mol e⁻ corresponds to 0.004634 mol H_2 evolved.
(0.004634 mol)(0.08206 L·atm/mol·K)(273K)/(1.00 atm) = **0.104 L H_2** evolved.
0.009268 mol e⁻ corresponds to 0.002317 mol O_2 evolved
(0.002317 mol)(0.08206 L·atm/mol·K)(273K)/(1.00 atm) = **0.0519 L O_2** evolved.

14.59 a. $4OH^- + VO^{2+} \rightarrow VO_3^- + 2H_2O + e^-$
b. $7H_2O + 2Cr^{3+} \rightarrow Cr_2O_7^{2-} + 14H^+ + 6e^-$
c. $2H_2O + Mn^{2+} \rightarrow MnO_2 + 4H^+ + 2e^-$
d. $4OH^- + NO \rightarrow NO_3^- + 2H_2O + 3e^c$
e. $Fe^{3+} + e^- \rightarrow Fe^{2+}$

CHAPTER FOURTEEN

14.61 Mn + 2AgCl → Mn^{2+} + 2Ag + Cl$^-$

E° = 1.18V − 0.222V = 0.958V

The anode will consist of a piece of manganese metal in contact with aqueous 1 M Mn^{2+}. The cathode will consist of a piece of silver metal in contact with a solid AgCl which is equilibrium with its ions. A salt bridge will connect the two solutions and a wire will connect the two electrodes. Electrons will flow from the manganese to the silver.

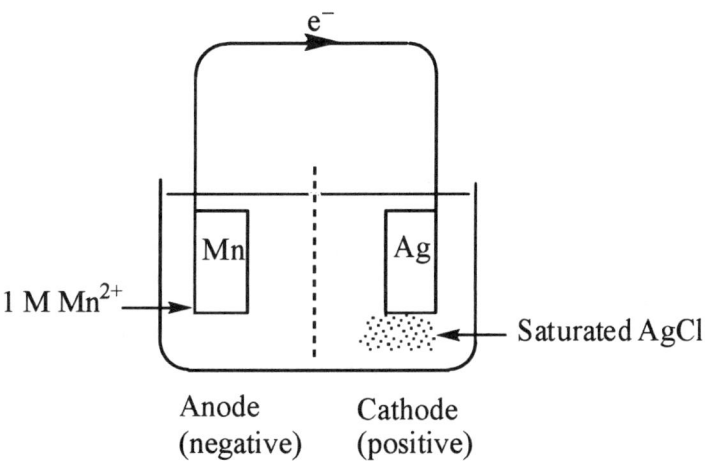

14.63 a. E = E° − (0.0591/n) log Q
E = −0.77 − (0.0591/1) log(2) = **−0.79 V**

b. E = E° − (0.0591/2) log Q
E = 0.250 − (0.0591/2) log(0.0010) = **0.338 V**

c. E = E° − (0.0591/5) log Q
E = 1.51 − (0.0591/5) log[(0.10)/(1.00)(0.1)^8] = **1.43 V**

14.65 We can calculate the equilibrium constant from the standard cell potential.
E° = 0.799 V − 0.788 V = 0.011 V
K = 10$^{(nE°/0.0591)}$ = **2.36**

The reaction would spontaneously proceed **to the right** with all substances in their standard states.

14.67 a. E° = 0.44 V − 0.277 V = **0.16 V**

E = E° − (0.0591/n) log Q = 0.16 V − (0.0591/2) log (1.0/0.50) = **0.15 V**

When E = 0, 0.16 V = (0.0591/2) log [Fe^{2+}]/[Co^{2+}] and
[Fe^{2+}]/[Co^{2+}] = 2.6 × 10^5

CHAPTER FOURTEEN

b. $E° = 0.037 \text{ V} + 1.679 \text{ V} = \textbf{1.707 V}$

$E = E° - (0.0591/n) \log Q$

$E = 1.707 \text{ V} - (0.0591/3) \log[(0.010)/[(0.10)^4(0.010)]] = \textbf{1.628 V}$

14.69 H_2 will be produced at the cathode and O_2 at the anode.

Twice as many moles of H_2 will be produced as O_2.

Cathode: $4H_2O + 4e^- \rightarrow 2H_2 + 4OH^-$, $E° = -0.828$ V

Anode: $2H_2O \rightarrow 4e^- + 4H^+ + O_2$, $E° = -1.23$ V

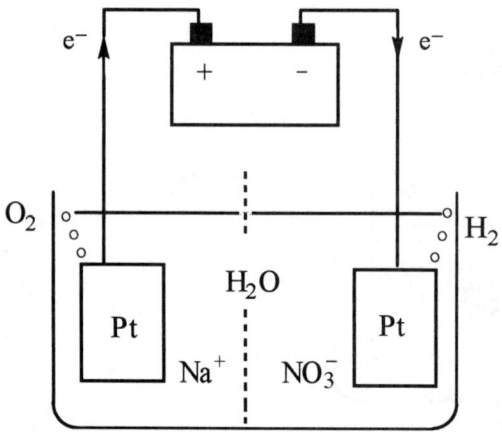

14.71 First, we must write the balanced equation for the process. In this reaction, Na_2SO_3 is oxidized to Na_2SO_4 and XeO_3 is reduced to Xe.

Reduction: $6H^+ + XeO_3 + 6e^- \rightarrow Xe + 3H_2O$

Oxidation: $H_2O + Na_2SO_3 \rightarrow Na_2SO_4 + 2e^- + 2H^+$

Overall: $XeO_3 + 3Na_2SO_3 \rightarrow Xe + 3Na_2SO_4$

We will find the number of moles of Xe produced and the mass of Na_2SO_3 required to produce this quantity of Xe. Then we will use this information to calculate the percentage of Na_2SO_3 in the original sample.

$n_{Xe} = PV/RT = (0.950 \text{ atm})(0.175 \text{ L})/(0.08206 \text{ L·atm/mol·K})(308.2\text{K})$

$n_{Xe} = 6.57 \times 10^{-3}$ mol Xe.

$(6.57 \times 10^{-3} \text{ mol Xe})(3 \text{ mol Na}_2SO_3/1 \text{ mol Xe}) = 0.0197$ mol Na_2SO_3

$(0.0197 \text{ mol Na}_2SO_3)(126.04 \text{ g/mol}) = 2.48$ g Na_2SO_3

Since there were 2.48 g Na_2SO_3 in the original 5.00-g sample, the percentage of Na_2SO_3 was (2.48 g/5.00 g) = 0.496 = **49.6%**.

CHAPTER FOURTEEN

14.73 The spontaneous cell reaction will have Zn being oxidized to Zn^{2+} and IO_3^- being reduced to I_2.

Oxidation: $Zn \rightarrow Zn^{2+} + 2e^-$

Reduction: $2IO_3^- + 12H^+ + 10e^- \rightarrow I_2 + 6H_2O$

Overall: $5Zn(s) + 2IO_3^-(aq) + 12H^+(aq) \rightarrow 5Zn^{2+}(aq) + I_2(s) + 6H_2O(liq)$

$E° = 0.763\ V + 1.19\ V = \mathbf{1.95\ V}$

In order to calculate the actual cell potential, we need to know the concentration of IO_3^- in 0.100 M HIO_3. If we let "x" moles per liter of HIO_3 dissociate in order to achieve equilibrium, then $[HIO_3] = (0.100 - x)$ and $[H^+] = [IO_3^-] = x$.

$K_a = 0.19 = x^2/(0.100 - x)$; $0.019 - 0.19x = x^2$

$x^2 + 0.19x - 0.019 = 0$

$x = [-0.19 \pm (0.19^2 + 0.076)^{1/2}]/2$

$x = 0.072$ (plus a negative root which will not yield a physically meaningful answer).

So $[IO_3^-] = 0.072$ M in the half-cell with the Pt electrode.

Now we use the Nernst equation to calculate the potential of the cell.

$E = E° - (0.0591/n) \log Q$

$E = 1.95\ V - (0.0591/10) \log[Zn^{2+}]^5/[H^+]^{12}[IO_3^-]^2$

$E = 1.95\ V - (0.0591/10) \log(0.100)^2/(0.072)^{12}(0.072)^2$

$E = 1.95\ V - 0.08\ V = \mathbf{1.87\ V}$

14.75 The mass of alumina present is $(1.0 \times 10^3\ kg)(0.015) = 15\ kg\ Al_2O_3$

Molar masses are Al_2O_3: 101.96 g/mol; Al: 26.98 g/mol

The amount of Al recovered =

$(15\ kg\ Al_2O_3)(1\ mol/101.96\ g/mol)(2\ mol\ Al/1\ mol\ Al_2O_3)(26.98\ g/mol) = \mathbf{7.9\ kg}$

14.77 The balanced equation for the electrolysis of molten sodium chloride is
$2NaCl \rightarrow 2Na + Cl_2$.

$(10.0 \times 10^3\ g\ NaCl)(1\ mol/58.44\ g) = 171\ mol\ NaCl$

$(171\ mol\ NaCl)(1\ mol\ Cl_2/2\ mol\ NaCl) = 342\ mol\ Cl_2$

Since the yield is 96.0%, we will only obtain $(0.960)(342\ mol) = 328\ mol\ Cl_2$.

Now calculate the pressure exerted by the Cl_2 under the given conditions by the Van der Waals equation:

$(P + n^2a/V^2)(V - nb) = nRT$. Solve for pressure:

$P = (nRT)/(V - nb) - n^2a/V^2$. The Van der Waals constants for Cl_2 are available from Table 4.4: $a = 6.493\ L^2 \cdot atm/mol^2$, $b = 0.05622\ L/mol$

$P = (328)(0.08206)(308)/(75.0 - 18.44) - 124.185$

$P = \mathbf{22.4\ atm}$

CHAPTER FOURTEEN

14.79 The balanced equation for the reaction of ammonia with oxygen to form nitrogen is:
$4NH_3 + 3O_2 \rightarrow 2N_2 + 6H_2O$

(100. g NH_3)(1 mol/17.03 g) = 5.87 mol NH_3

(5.87 mol NH_3)(12 mol e^-/4 mol NH_3)(96,500 coul/mol e^-) = 1.70×10^6 coul

Since this cell operates at 40.0% efficiency, only (0.400)(1.70×10^6 coul)
= **6.80×10^5 C** will be produced.

14.85 a. The atomic mass of Cd is more than double that of Fe (112.4 g/mol vs. 55.8 g/mol). Starting with equal masses of each, there will be fewer than half the moles of Cd compared to Fe. With equal currents and times, a cadmium battery will undergo a greater extent of chemical change.
b. The cadmium battery, having fewer total moles of reactants, will have less chemical energy to convert to electrical energy.
c. The cadmium battery, having fewer total moles of reactants, will both charge and discharge faster than the Fe battery, assuming equal charge/discharge currents were used in both batteries

CHAPTER FIFTEEN

Chemical Kinetics

15.1 Reactions b and c are not likely to be elementary processes, since they involve 4 particles reacting at once. There are no known molecular processes greater than termolecular (3 particles simultaneously reacting).

15.3   a. $-\Delta[A_2]/\Delta t = -\Delta[B_2]/\Delta t = (1/2)\,\Delta[AB]/\Delta t$
     b. $-\Delta[A_2]/\Delta t = -\Delta[B_2]/\Delta t = \Delta[A_2B_2]/\Delta t$

15.5   a. Since the reaction mechanism involves a single, elementary step, we can write the rate law from the stoichiometric coefficients: Rate = $k[NO]^2[Cl_2]$
The reaction is **third order** overall, first order in Cl_2 and second order in NO.
     b. $(1/2)\,\Delta[NOCl]/\Delta t = -\Delta[Cl_2]/\Delta t$

15.7 The reaction is first order in H_3AsO_4, first order in I^- and second order in H_3O^+.
Rate = $k[H_3AsO_4][H_3O^+]^2[I^-]$
$3.70 \times 10^5 = k(0.001)(0.010)^2(0.10)$
$k = 3.7 \times 10^{13}$ $L^3/mol^3 \cdot s$

So the correct experimental rate law is:
Rate = $-\Delta[H_3AsO_4]/\Delta t = (3.7 \times 10^{13}\ L^3/mol^3 \cdot s)[H_3AsO_2][H_3O^+]^2[I^-]$

15.9   a. In order to determine the overall order of the reaction, we need to determine the order with respect to each component. Since doubling [A] doubles the rate, the reaction is first order in [A]. Since doubling [B] quadruples the rate, the reaction is second order in [B], so the reaction is third order overall.
     b. In part (a), we determined that the reaction is first order in [A] and second order in [B].
     c. The rate law can be written rate = $k[A][B]^2$. Inserting the data from experiment #1, we get: $(5 \times 10^{-4}\ M/min) = k(0.500\ M)(0.050\ M)^2$.
$k = 0.4\ M^{-2}\ min^{-1}$

15.11 For a first-order process, $t_{1/2} = 0.693/k = 0.693/(3.5 \times 10^{-5}\ s^{-1})$.
$t_{1/2} = \mathbf{2.0 \times 10^4\ s}$

15.13 First determine the rate constant from $t_{1/2}$:
$k = 0.693/t_{1/2} = 0.693/10.\ min = 6.93 \times 10^{-2}\ min^{-1}$
Then determine the time for the reaction to be 99% complete: $[A]/[A]_o = 0.01$
$\ln [A]/[A]_o = -kt = \ln 0.01$
$t = -(1/6.93 \times 10^{-2}\ min^{-1})(-4.60) = \mathbf{66\ min}$

CHAPTER FIFTEEN

15.15 We will use the integrated form of the rate law for a first order reaction: $\ln([A]_o/[A]) = kt$. The half-life for a first-order process is $t_{1/2} = 0.693/k$.

In each minute, the concentration of A decreases by 0.01. Thus, after one minute of change, we have $[A]/[A]_o = 0.99$, and $[A]_o/[A] = 1.01$. Then
$k = \ln(1.01)/1$ min $= 0.010$ min^{-1}, and **$t_{1/2} = 0.693/k = 69.3$ min.**

15.17 a. The rate law for a first order process is $-\Delta[A]/\Delta t = k[A]$.

b. Let $t_{90\%}$ = time for reaction to go to 90% completion.
When the reaction is 90% complete, $[A]/[A]_o = 0.10$ and $[A]_o/[A] = 10$.
$\ln(10) = kt_{90\%} = 2.303$
$t_{90\%} = 2.303/k$

15.19 $k = Ae^{-E_a/RT}$
$= (1.3 \times 10^{11}$ s$^{-1})e^{-5.5 \times 10^4 \text{ J/mol}/[(8.314 \text{ J/mol·K})(298K)]} =$ **30. s^{-1}**

15.21 We know that $\ln(k_2/k_1) = (-E_a/R)[1/T_2 - 1/T_1]$.
$\ln(576/47.5) = (-E_a/8.314 \times 10^{-3}$ kJ/mol·K$)$ $[1/313K - 1/293K]$
$E_a = 95$ kJ/mol
At 303K, $\ln(k/47.5 \times 10^{-5}$ s$^{-1}) = [(-95)/(8.314 \times 10^{-3})][1/303 - 1/293]$
$k = 1.72 \times 10^{-3}$ s^{-1}. For a first order reaction, we know that $t_{1/2} = 0.693/k$.
$t_{1/2} = 0.693/(1.72 \times 10^{-3}$ s$^{-1}) =$ **403 s**

15.23 $\ln(k_2/k_1) = (-E_a/R)[1/T_2 - 1/T_1]$
$\ln(575/45.0) = (-E_a/0.008314$ kJ/mol·K$)[1/313K - 1/293K]$
$E_a = 97$ kJ/mol
At 303K, $\ln(k/45.0 \times 10^{-5}$ min$^{-1}) = (-97/0.008314)[1/303 - 1/293]$
$k = 1.67 \times 10^{-3}$ min^{-1}

15.25 a. There are two elementary steps:
A + B → C (slow) followed by
C → P

b. The rate of the overall reaction will be the rate of the slow step:
Rate = $k_1[A][B]$.

CHAPTER FIFTEEN

15.27 One possibility would be:
$NO_2 + F_2 \rightarrow NO_2F + F$ (slow)
$F + NO_2 \rightarrow NO_2F$ (fast)
If the first step is rate-determining, then
Rate = $k_1[NO_2][F_2]$
Since NO_2 and F are both radicals (that is, they contain unpaired electrons), they are both very reactive. Each species may recombine with itself rather than combining with each other in the second step of the proposed mechanism.
$F + F \rightarrow F_2$; $NO_2 + NO_2 \rightarrow N_2O_4$
Each of these side reactions would complicate the proposed mechanism and slow down the overall reaction.

15.29 a. For a second order process, Rate = $k[A]^2$.
k = Rate/$[A]^2$ = (M/s)/(M)^2 = **M^{-1} s^{-1}**.
b. The initial rate is $k[Ti^{3+}]^2$ = $(1.0 \times 10^2 \, M^{-1} \, s^{-1})(0.015)^2$
Rate = 0.022 M/s
c. Rate = $-1/2 \, \Delta[Ti^{3+}]/\Delta t$ = 0.022 M/s
$\Delta[Ti^{3+}]/\Delta t = -2(0.022 \, M/s)$
= **−0.044 M/s**
d. For a second order process,
$1/[A] - 1/[A]_o = kt$
We want the time for [A] to be 1/10 $[A]_o$ or for $[A]_o/[A]$ = 10. Multiply both sides of the second order equation by $[A]_o$:
$[A]_o/[A] - 1 = [A]_o \, kt$
$10. - 1 = (0.015 \, M)(1.0 \times 10^2 \, M^{-1} \, s^{-1})t$
t = 6 s

15.31 a. Since log[C] vs. time is a straight-line graph, we know that the reaction is **first order in [C]**.
b. We know that $\ln([C]_o/[C]) = kt$, so $2.303 \log([C]_o/[C]) = kt$.
k = (2.303/70 min) (log$[C]_o$ − log[C]) = (2.303/70 min) (0.4 + 0.8)
k = 0.039 min^{-1}

15.33 a. We know that $\ln([A]/[A]_o) = -kt$.
$\ln([A]/[A]_o) = -(0.0123 \, min^{-1})(60 \, min) = -0.738$
$[A]/[A]_o = 0.478$, so 52.2% of the original ethylene oxide has decomposed to products after 60 min at 415°C.
b. When 75% of the starting material has reacted, $[A]/[A]_o$ = 0.25.
$\ln([A]_o/[A]) = 1.39 = (0.0123 \, min^{-1})t$; **t = 113 min**

CHAPTER FIFTEEN

15.35 a. Elementary processes higher than third order are not known It is uncommon to have three particles collide simultaneously, and more than that are unknown. Reactions that involve multiple reactants usually proceed via multi-step mechanisms.
b. If this reaction were elementary, it would be first order in [A] and fourth order overall. The rate law would be $\Delta[X]/\Delta t = k[A][B][C][D]$.

15.37 There are several propagation steps. One of them is $Cl\cdot + CH_4 \rightarrow CH_3Cl + H\cdot$.

A likely transition state would involve carbon having five bonds. This is obviously not a stable situation, and the transition state will decompose into CH_3Cl and a hydrogen atom.

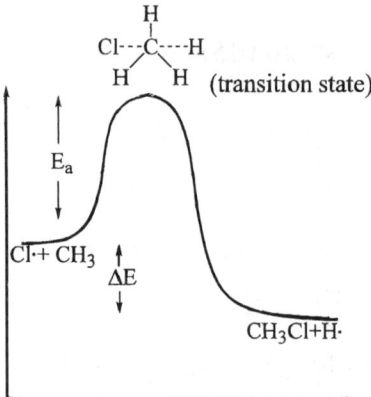

15.39 The catalyst species is Mn^{2+}. The rate law is
$-\Delta[Fe^{3+}]/\Delta t = k[Fe^{3+}][Mn^{2+}]$.

15.41 $A(g) + B(g) \rightarrow C(g) + D(g)$

$K = p_C p_D / p_A p_B = k_f / k_r = (2.44 \times 10^{-7})/(3.19 \times 10^{-1}) = 7.65 \times 10^{-7}$

If we start with 0.10 atm A and 0.15 atm B and we let the reaction occur by an amount "x" in order to achieve equilibrium, then $p_A = (0.10 - x)$, $p_B = (0.15 - x)$ and $p_C = p_D = x$.

$K = 7.65 \times 10^{-7} = x^2/[(0.10 - x)(0.15 - x)] \approx x^2/[(0.10)(0.15)]$

$x^2 \approx (0.10)(0.15)(7.65 \times 10^{-7}) = 1.15 \times 10^{-8}$

$x \approx 1.07 \times 10^{-4}$ (approximation valid)

At equilibrium, **$p_A = 0.10$ atm, $p_B = 0.15$ atm, $p_C = p_D = 1.07 \times 10^{-4}$ atm.**

CHAPTER FIFTEEN

15.45 a. A first order process has a half-life that is independent of the starting concentration or pressure: $t_{1/2} = 0.693/k$. For a second order process, $t_{1/2}$ varies inversely with $[A]_o$: $t_{1/2} = 1/k[A]_o$. From the data given, we see that $t_{1/2}$ becomes shorter as $[A]_o$ (expressed in torr) increases. Of the two choices, the reaction must be **second order**.

b. First, determine the rate constant in torr units from either of the initial pressures:
$k = 1/(t_{1/2} P^o) = 1/(250 \text{ s})(300. \text{ torr})$
$= 1.33 \times 10^{-5} \text{ torr}^{-1}\text{s}^{-1}$

(the other set of data will give the same value for k).

Now determine the half-life starting from an initial pressure of 1 atm = 760 torr:
$t_{1/2} = 1/kP^o = 1/[(1.33 \times 10^{-5} \text{ torr}^{-1} \text{ s}^{-1})(760 \text{ torr})]$
$= \mathbf{98.7 \text{ s}}$ (or $\mathbf{1 \times 10^2 \text{ s}}$ if 1 atm is meant to have one significant figure)

CHAPTER SIXTEEN

Solids

16.1 For the face-centered cubic structure, $a = 2\sqrt{2}\cdot r$ and $V = a^3$, so $V = (2\sqrt{2}\cdot r)^3$.
Doubling r would result in 8 times the original volume and, since $d = m/V$, 1/8 the original density.

16.3 The density of the metal must equal the density of the unit cell. For tungsten, each unit cell contains 2 tungsten atoms and has a mass of $2(183.85 \text{ g/mol})/(N_A)$. The volume of the unit cell is $(3.1583 \times 10^{-8} \text{ cm})^3$.

    Density = m/V
    $19.3 \text{ g/cm}^3 = [2(183.85 \text{ g/mol})/(N_A)]/(3.153 \times 10^{-8} \text{ cm})^3$
    $N_A = 2(183.85 \text{ g/mol})/[(19.3 \text{ g/cm}^3)(3.153 \times 10^{-8} \text{ cm})^3]$
    $N_A = 6.08 \times 10^{23}$

16.5 The density of the metal must equal the density of the unit cell. For silver, each unit cell contains 4 Ag atoms and has a mass of $4(M_{Ag})/N_A$. The volume of the unit cell is $(4.0776 \times 10^{-8} \text{ cm})^3$.

    $10.5 \text{ g/cm}^3 = 4(M_{Ag})/[(6.022 \times 10^{23}/\text{mol})(4.0776 \times 10^{-8} \text{ cm})^3]$.
    $M_{Ag} = 107 \text{ g/mol}$

16.7 The density of the metal must equal the density of the unit cell. For copper, each unit cell contains 4 copper atoms and has a mass of $4(M_{Cu})/N_A$. The volume of the unit cell is $(3.61 \times 10^{-8} \text{ cm})^3$, and density = m/V.

    $d = 4(63.546 \text{ g/mol})/[(6.022 \times 10^{23}/\text{mol})(3.61 \times 10^{-8} \text{ cm})^3]$
    $d =$ **8.97 g/cm^3**
This value of the density is in good agreement with the given value of 8.92 g/cm^3.

16.9 In the face-centered cubic lattice of calcium, the face-diagonal of the unit cell is equal to 4r, where r = the atomic radius of calcium. The edge length of the unit cell, a, is equal to $(2\sqrt{2})r$.
    $a = 5.56\text{Å} = (2\sqrt{2})r$; r = **1.96 Å**

16.11 The density of the copper must be the same as the density of its unit cell.
In the face-centered cubic lattice, there are four atoms per unit cell and each unit cell has a mass of $4(\text{atomic weight})/N_A$.
    $8.92 \text{ g/cm}^3 = m/V$
        $m = [4(63.55 \text{ g/mol})/(6.022 \times 10^{23}/\text{mol})] = (4.221 \times 10^{-22} \text{ g})$
        $V = m/d = (4.221 \times 10^{-22} \text{ g})/(8.92 \text{ g/cm}^3)$

CHAPTER SIXTEEN

$V = 4.73 \times 10^{-23}$ cm^3 = a^3, where a = edge length of the unit cell

a = 3.62×10^{-8} cm = 3.62 Å

In the face-centered cubic unit cell of copper, the face diagonal = 4r, where r = atomic radius of copper. We know that the edge length of this cube is 3.62 Å which means that the face-diagonal is equal to ($\sqrt{2}$)(3.62 Å) = 5.12 Å.

5.12 Å = 4r; **r = 1.28 Å**

16.13 A free metal in the fcc structure has a unit cell that contains 4 metal atoms. If r = atomic radius, then the total atomic volume/unit cell = 4(4/3 πr^3) = 16.76 r^3. The fcc unit cell has a = (2$\sqrt{2}$)r and a total volume of [2($\sqrt{2}$)r]^3 = 22.63 r^3. Thus, (22.63 – 16.76)r^3 is the volume of empty space. The fraction of empty space is 5.87r^3/22.63r^3 = 25.9%.

16.15 BaCl$_2$ has the fluorite structure, with Ba^{2+} and Cl$^-$ ions in contact. Therefore the distance between ion centers is 1.29 Å + 1.81 Å = **3.10 Å**.

16.17 For the CsCl structure, we know that the sum of the ionic radii is equal to (a$\sqrt{3}$)/2.

$r_{Br^-} + r_{Tl^+}$ = (a$\sqrt{3}$)/2 = ($\sqrt{3}$) (3.97 Å)/2 = 3.44 Å

r_{Tl^+} = 3.44 Å – 1.82 Å = **1.62 Å**

16.19 The density of the compound must be the same as the density of the unit cell. In the sodium chloride structure, each unit cell contains 4 Na$^+$ and 4 Cl$^-$. It has a mass of 4(formula weight)/N$_A$ and a volume of [2(2.83 Å)]^3,

d = 2.167 g/cm^3 = m/V

= [4(58.443 g/mol)/(N$_o$)]/(5.66 × 10^{-8} cm)^3

2.167 g/cm^3 = (233.772 g/mol)/N$_A$/1.813 × 10^{-22} cm^3

N$_A$ = (233.772 g/mol)/[(2.167 g/cm^3)(1.813 × 10^{-22} cm^3)

N$_A$ = 5.95 × 10^{23}

16.21 In the diamond structure, there are 8 total atoms per unit cell; 8 corners (1/8 atom each), 6 faces (1/2 atom each) and 4 interior atoms. So there are **eight** silicon atoms per unit cell.

16.23 In the ionic FCC structure, the unit cell length is

a = 2 $r_{Ba^{2+}}$ + 2 $r_{S^{2-}}$

= 2(1.94 Å) + 2(1.90 Å) = **6.38 Å**

CHAPTER SIXTEEN

16.25 The density of the compound must be the same as the density of the unit cell. Since LiCl is stated to have the sodium chloride structure, each unit cell contains 4 Li^+ and 4 Cl^-. It has a mass of 4(formula weight)/N_A and a volume of $[2(2.57 \text{ Å})]^3$.

 d = m/V
 m = 4(42.39 g/mol)/(6.022 × 10^{23}/mol) = 2.816 × 10^{-22} g
 V = (5.14 × 10^{-8} cm)^3 = 1.36 × 10^{-22} cm^3
 d = 2.816 × 10^{-22} g/1.36 × 10^{-22} cm^3 = **2.07 g/cm^3**

16.27 The density of the substance must be the same as the density of the unit cell. Since the unit cell contains four molecules, its mass is 4(molecular weight)/N_A. Its volume is (5.15 × 10^{-8} cm)^3 = 1.37 × 10^{-22} cm^3.

 d = m/V
 0.73 g/cm^3 = [4(M)/(6.022 × 10^{23}/mol)]/(1.37 × 10^{-22} cm^3)
 M = 15.0 g/mol

As a gas at 25°C and 1.00 atm, the density of the substance can be calculated from the ideal gas equation:

PV = nRT = (m/M)RT; d = m/V = PM/RT
d = (1.00 atm)(15.0 g/mol)/(0.08206 L·atm/mol·K)(298K)
 = **0.613 g/L**

16.29 We will consider the unit cell for graphite to be a hexagonal prism 3.35 Å in height. From the density of graphite, we will be able to calculate the area of the hexagonal base and then the carbon-carbon bond distance within the planes of carbon atoms. Each carbon atom is shared by six unit cells so in each unit cell we have 12(1/6) = 2 carbon atoms. The density of the graphite must be the same as the density of the unit cell. The mass of the unit cell is 2(12.011 g/mol) / (6.022 × 10^{23}/mol) = 3.989 × 10^{-23} g, and its volume is (h.a.)(3.35 Å) where h.a. = hexagonal area of the base of the unit cell.

 d = 2.25 g/cm^3 = m/V
 2.25 g/cm^3 = (3.989 × 10^{-23} g) / (h.a.)(3.35 × 10^{-8} cm)
 h.a. = 5.29 × 10^{-16} cm^2 = 5.29 Å^2

If we let x = the carbon-carbon bond distance within the planes of carbon atoms, then, then "x" is also equal to the edge length of the hexagon. The hexagon can be broken into six equilateral triangles, each one having base = x and height = ($\sqrt{3}$)x/2. The area of each triangle is (1/2)(x)($\sqrt{3}$)(x/2), and the area of the hexagon = h.a. = 6(1/2)(x)($\sqrt{3}$)(x/2) = (3$\sqrt{3}$)x^2/2.
5.29 Å^2 = (3$\sqrt{3}$)x^2/2; x = 1.43 Å

CHAPTER SIXTEEN

The average carbon-carbon bond length for a single bond is 1.54 Å and the average bond length for a double bond is 1.34 Å. The calculated bond length in graphite is 1.43 Å which corresponds roughly to a 3/2 bond length. The carbon atoms in graphite must be experiencing resonance of the same type that is characteristic of benzene (see Chapter 7). The carbon atoms are sp^2-hybridized, with delocalized electrons occupying parallel p-orbitals on carbon atoms within a plane. There is little bonding between planes of carbon in graphite, which is consistent with the large difference in distance between planes as opposed to carbon-carbon bond distances within a plane (3.35 Å vs. 1.43 Å).

16.31 One gram of silicon contains (1 g)(1 mol/28.0855 g)(6.022×10^{23}/mol) = 2.144×10^{22} atoms of silicon. At 99.9999999% purity, we have 0.0000001% impurity. $(1 \times 10^{-9})(2.144 \times 10^{22}$ atoms) = **2.144×10^{13} atoms of silicon** have been replaced by some other element.

16.39 a. The face-centered orthorhombic unit cell has eight corner points and six end (face) points.
(8 corner)(1/8 point/corner) + (6 faces)(1/2 point/face) = **4 lattice points**
b. The primitive triclinic unit cell has eight corner points.
(8 corners)(1/8 point/corner) = **1 lattice point**
c. The body-centered monoclinic unit cell has eight corner points and one center point.
(8 corner)(1/8 point/corner) + (1 center)(1 point/center) = **2 lattice points**

16.41 $Fe_{0.97}O_{1.00}$ contains both Fe^{2+} and Fe^{3+} ions. The oxygen ions are all O^{2-}, so the average charge to balance this must be +2. Let x = amount of Fe^{3+}; therefore, the amount of Fe^{3+} will be 0.97 − x.
(x)(+3) + (0.97 − x)(+2) = 2.00 ∴ x = 0.060
The fraction of Fe ions in the +3 state is 0.060/0.97 = 0.062 or 6.2%

16.43 $Ni_{0.97}O_{1.00}$ contains both Ni^{2+} and Ni^{3+} ions. The oxygen ions are all O^{2-}, so the average charge to balance this must be +2. Let x = amount of Ni^{2+}; therefore, the amount of Ni^{3+} will be 0.97 − x.
(x)(+2) + (0.97 − x)(+3) = 2.00 ∴ x = 0.910
The fraction of Ni ions in the +2 state is 0.910/0.97 = 0.94 or 94%

CHAPTER SEVENTEEN

Materials

17.1 a. For a paperclip, the material should be flexible, strong, inexpensive and shatter-resistant. A simple metal or a polymer would be appropriate.
  b. For a compact disk, the material should be flexible, inexpensive and shatter-resistant and it should not creep over time. A simple metal or polymer would be appropriate although a composite with minimum tendency to creep would be even better.
  c. A raincoat should be flexible, inexpensive, moisture-resistant and attractive. A polymer or composite material would be appropriate.
  d. A chair should be strong, flexible and resistant to wear. Chairs for different purposes will have different additional requirements. A variety of materials (metals, polymers, composites) could be, and have been, used in the production of chairs.
  e. For an insulator for a high voltage wire, the material should be strong, flexible, non-conducting and resistant to heat. A structural ceramic, a heat-resistant polymer or a composite material would be appropriate.

17.3 Van der Waals forces between the molecules will break first; they correspond to intermolecular forces and are weaker than formal covalent bonds.

17.5 a. SiC forms a covalent lattice similar to that of diamond (which melts at around 3500°C, Table 17.4). SiC should melt at a temperature somewhat less than 3500°C. Its actual melting point is around 2700°C.
  b. MgO should have an ionic structure similar to CaO (which melts at 2580°C, Table 17.4). MgO should melt at a somewhat higher temperature due to the smaller size of the Mg^{2+} ion as compared with the Ca^{2+} ion. Its actual melting point is 2800°C.
  c. Cu is held together by metallic bonding and is close to Fe on the periodic table (Fe melts at 1535°C, Table 17.4). Cu should melt at a temperature not very different from that of Fe. The actual melting point of copper is 1083°C.

17.7 a. 1.00 mm diameter = 1.00×10^{-3} m.
  The cross sectional area A = $\pi r^2 = \pi (5.00 \times 10^{-4} \text{ m})^2 = 7.85 \times 10^{-7} \text{ m}^2$
  Stress = F/A = $(50.0 \text{ kg})(9.81 \text{ m/s}^2) / 7.85 \times 10^{-7} \text{ m}^2 = \mathbf{6.25 \times 10^8 \text{ N/m}^2}$
  b. Strain = length increase/original length
  = (51.3 cm − 50.0 cm)/50.0 cm = **2.6×10^{-2} cm/cm**

CHAPTER SEVENTEEN

17.9 Table 17.6 shows that Be has a hcp structure in common with Cd, Co, Mg and Zn. Since these metals all have the same crystal structure, the ratio of their atomic radii should be the same as the ratio of their unit cell lengths. Percent differences in atomic radii are as follows:

Cd: $100\% \times |2.973 - 2.283|/2.283 = 30\%$
$100\% \times |5.606 - 3.607|/3.607 = 55\%$

Co: $100\% \times |2.514 - 2.283|/2.283 = 10\%$
$100\% \times |4.105 - 3.607|/3.607 = 14\%$

Mg: $100\% \times |3.203 - 2.283|/2.283 = 40\%$
$100\% \times |5.196 - 3.607|/3.607 = 44\%$

Zn: $100\% \times |2.658 - 2.283|/2.283 = 16\%$
$100\% \times |4.934 - 3.607|/3.607 = 37\%$

Only Co falls within the atomic radius requirement of the Hume-Rothery rules. The electronegativities of Be and Co are 1.5 and 1.8 which can be considered similar. Therefore, **cobalt** would be expected to form solid solutions with beryllium.

17.11 a. A **liquid solution** (70% Ag, 30% Cu) exists under these conditions.
b. A **heterogeneous mixture** of a **solid solution** (containing roughly 95% Ag, 5% Cu) and a **liquid solution** (containing roughly 80% Ag, 20% Cu) exists under these conditions.

17.13 5 SiO_4 units would have 5 Si atoms and 20 oxygen atoms with a total charge of –20. Four oxygen atoms an 8 negative charges are removed upon linking, so the formula would be $Si_5O_{16}^{12-}$

17.15 Representative structures are:
a.

$Si_8O_{24}^{16-}$

CHAPTER SEVENTEEN

b. The infinite sheet structure can be drawn as follows:

$(Si_2O_5^{2-})_n$

An alternative structure could be:

$(Si_2O_5^{2-})_n$

17.17 Aluminosilicate ions have the general formula $Si_mAl_nO_{2(m+n)}{}^{n-}$.
   a. $m = 3$; $n = 1$; $2(m + n) = 8$; $y = n = 1$
      The correct formula is **$Si_3AlO_8^-$**
   b. $m = 2$; $2(m + n) = 8$; so $x = y = n = 2$
      The correct formula is **$Si_2Al_2O_8^{2-}$**

17.19 The atomic packing factor, APF, is equal to the ratio of atomic volume to total volume in the unit cell of a metal. It is the fraction of space occupied by actual atoms in the metal. The face-centered cubic structure contains eight corner atoms and six atoms on faces of the cube.
(8 corners)(1/8 atom/corner) + (6 faces)(1/2 atom/face)
The total atomic volume within the unit cell is $4(4/3\pi r^3) = 16.76\, r^3$. The face diagonal of a face-centered cubic structure is equal to 4r. This means that the edge length a, is equal to $2\sqrt{2}\, r$. The total unit cell volume is equal to $a^3 = (2\sqrt{2}\, r)^3 = 22.63\, r^3$.
APF = $(16.76\, r^3)/(22.63\, r^3)$ = **0.74**

CHAPTER SEVENTEEN

17.21 $-(CH_2-CH_2)_n- + 3nO_2 \rightarrow 2n\ CO_2 + 2n\ H_2O$

17.23 For every 100.0 g of the sample, we will have 55.0 g of Ag_2O and 45.0 g CuO. Reaction with hydrogen to yield pure metals will produce:
(55.0 g Ag_2O)(1 mol Ag_2O/231.74 g Ag_2O)(2 Ag/1 Ag_2O)(107.87 g Ag/1 mol Ag)
= 51.2 g Ag
(45.0 g CuO)(1 mol CuO/79.55 g CuO)(1 mol Cu/1 mol CuO)(63.55 g Cu/1 mol Cu)
= 35.9 g Cu
The liquid-liquid solution that results when these metals are heated will contain:
(51.2 g)/(51.2 g + 35.9 g) = 0.588 = 58.8% Ag
(35.9 g)/(87.1 g) = 0.1412 = 41.2% Cu
Using Figure 17.21, we determine that the solution will begin to solidify at approximately **850°C**.

Once the solution cools to 700°C, a heterogeneous mixture of a **silver-rich solid solution** (containing roughly 93% Ag, 7% Cu) and a **copper-rich solid solution** (containing roughly 7% Ag, 93% Cu) will exist.

17.25 Strain is equal to the increase in length of the wire divided by the original length of the wire.
ε = strain = 4.35×10^{-2} = **0.0435**

The modulus of elasticity can be calculated from the relationship E = τ/ε, where E = modulus of elasticity, τ = stress and ε = strain. From Problem 17.20, we know that the stress on this wire is equal to 4.52×10^5 psi.
E = τ/ε = $(4.52 \times 10^5$ psi$)/(0.0435)$ = **1.0×10^7 psi**

17.27 As the copper is hammered, it will be work hardened, pushing the defects to the edges of the crystal and effectively removing them. This will continue to happen as the copper is flattened. Some defects may, in fact, be introduced by the hammering, but the total number of flaws should decrease.

CHAPTER SEVENTEEN

17.35 a. Desired characteristics: Impact resistance, corrosion resistance, light weight. Choices: metal, polymer, composite. Metal pros: light weight, impact resistance. Cons: impact resistance compromised by denting, possible corrosion issues. Polymer pros: Light weight, corrosion resistance. Cons: possible compromise in impact resistance. Composite pros: light weight, corrosion resistance, impact resistance. Cons: must be non brittle for best impact resistance.

     b. Desired characteristics: High tensile strength, puncture resistance. Choices: Metal, polymer, composite. Metal pros: high tensile strength, puncture resistance. Cons: possible heavy weight. Polymer pros: puncture resistance, high tensile strength, light weight. Cons: very few polymers have high enough tensile strength for this application. Composite Pros: light weight, very high tensile strength. Cons: possibly brittle, compromising puncture resistance.

     c. Desired characteristics: Heat resistance, high melting point and limited electrical conductivity. Choices: metals and ceramics. Metals pros: heat resistance, electrical conductivity. Cons: needs very high melting point to withstand application, corrosion resistance at high temperatures. Ceramics pros: heat resistance, very high melting point. Cons: difficult to make electrically conductive, brittle, can crack from thermal shock.

     d. Desired characteristics: Extreme heat resistance, ability to insulate heat. Choices: Metals and ceramics. Metals pros: easily worked and fabricated. Cons: very few metals are both inexpensive and have a high enough melting point to handle molten iron. Metals are also poor thermal insulators. Ceramics pros: excellent heat resistance, excellent heat insulators. Cons: most ceramics are brittle.

17.37 The repeated bending and subsequent breakage of the tin can lid is due to the breaking of the metal-metal bonds, which cannot be reformed easily. When polyproplyene is bent, the long polymer molecules are not broken into smaller chains. Rather, the polymer chains slide in relation to one another with the overcoming of intermolecular Van der Waals forces. The item returns to its original shape upon releasing the stress, as the Van der Waals forces are re-established.

CHAPTER EIGHTEEN

Properties of Polymers

18.1 Representative answers: Synthetic: nylon windbreaker; partial: cotton blend T-shirt; natural: wool sweater.

18.3 Representative answers: rubber mat, plastic wrap, nylon serving spoon, plastic forks.

18.5 Representative answers: Vinyl siding, garden hose, roofing tar, exterior paint, swimming pool liner.

18.7 Representative answers: Steering wheel, seat fabric, dash panel, air bag, seat belt, floor mat.

18.9 Degree of polymerization = number of monomer units incorporated into each polymer molecule. If the average molar mass of a polymer is 13,800 and the molar mass of a monomer unit is 92, the degree of polymerization is:
DP = 13800/92 = **150**

18.11 M_n = (120000 + 135000 + 150000 + 175000 + 190000)/5
M_n = 154,000
M_w = [(120000^2 + 135000^2 + 150000^2 + 175000^2 + 190000^2)/(120000 + 135000 + 150000 + 175000 + 190000)]
M_w = 158,000

18.13 Polydispersity = M_w/M_n = 15800/15400 = **1.03**

18.15 If the polydispersity = 1.0, then M_n/M_w = 1.0; therefore **M_n = M_w**

18.17 Molar mass = degree of polymerization × molar mass of monomer
M = 180 × 10^8 = **1.9 × 10^4**

18.19 Degree of polymerization = number of monomer units incorporated into each polymer molecule. The average molar mass of a polymer is 113,000 and the formula of the repeating unit is -[NH-(CH$_2$)$_6$-CO]$_n$-, so the molar mass of a monomer unit is 127. The degree of polymerization is 86,400/127 = **890**.

CHAPTER EIGHTEEN

18.21 Use the definitions of the two average molar masses:

Number average = $M_n = \dfrac{\Sigma N_i M_i}{\Sigma N_i}$ and mass average = $M_w = \dfrac{\Sigma N_i M_i^2}{\Sigma N_i M_i}$.

a. Since the M_n values of the two polymers A and B are the same, we can obtain equal masses of each by mixing equal numbers of formula units of A and B. This number will be designated simply as "N". Then for the two components we can write:
$\Sigma N_A M_A = 25{,}000N$ and $\Sigma N_B M_B = 25{,}000N$
When N formula units of each are mixed, the resulting number-average molar mass is

$M_n = \dfrac{\Sigma N_A M_A + \Sigma N_B M_B}{N_A + N_B} = \dfrac{25{,}000N + 25{,}000N}{N + N} = 25{,}000$. Thus, not surprisingly, the **number-average molar mass of the mixture is also 25,000.**

b. Again, we will mix equal numbers of formula units of the two polymers.

For polymer A: $M_A = \dfrac{\Sigma N_A M_A^2}{\Sigma N_A M_A}$ and for polymer B: $M_B = \dfrac{\Sigma N_B M_B^2}{\Sigma N_B M_B}$

From part (a), we know that $\Sigma N_A M_A = 25{,}000N$ and $\Sigma N_B M_B = 25{,}000N$.
Then $\Sigma N_A M^2_A = (75)(25{,}000N)$ and $\Sigma N_B M^2_B = (225)(25{,}000N)$. Now put both sets of numbers into the equation for M_w for the mixture:

$M_w = \dfrac{\Sigma N_A M_A^2 + \Sigma N_B M_B^2}{\Sigma N_A M_A + \Sigma N_B M_B} = \dfrac{(75)(25{,}000N) + (225)(25{,}000N)}{25{,}000N + 25{,}000N}$

$= \dfrac{75{,}000 + 225{,}000}{2} =$ **150,000**

18.23 Propylene is a thermoplastic (straight chain polymer)
Polyisoprene is a thermoplastic elastomer (polymer with some crosslinking)

18.25 a. **Ethylene glycol**, since it is a difunctional molecule and would form straight chain condensation polymers:
n HOCH$_2$CH$_2$OH → -[OCH$_2$CH$_2$O]$_n$- + n H$_2$O
Such a polymer would have a large glass transition temperature region and would be a thermoplastic material.

b. **Glycerol**, since it is a trifunctional molecule. It would form condensation polymers in the same manner as ethylene glycol, but since it is trifunctional, it would form crosslinked networks rather than linear chains. This would result in a three dimensional structure, which would have a much higher and more abrupt glass transition temperature. Furthermore, even if the degree of initial crosslinking were not great, additional heating cycles would result in further crosslinking and thermosetting.

CHAPTER EIGHTEEN

18.27 A plasticized polymer would have more entropy than a pure polymer, since to plasticize a polymer, additives are introduced to give the overall material different properties. Mixtures have higher entropies than the pure substances that comprise them.

18.29 Polymeric materials have more individual bonds; side chains can rotate, vibrate, stretch, all of which are means by which it can absorb energy.

18.33 It is possible to crystallize some polymers and check for regular, long range structure by X-ray diffraction. Also, IR spectroscopy confirms that the covalent bonds in a polymer backbone are not only similar, but equivalent.

18.35 The carbon-carbon bonds in the backbone of polyisoprene alternate -[sp^2-sp^3-sp^3-sp^2]$_n$-, so the bond angles between successive carbon atoms in the backbone would be 109.5° between sp^2-sp^2 carbon atoms and 120° between sp^3-sp^3 carbon atoms. Natural branching involves the breaking of carbon-carbon double bonds to make carbon-carbon single bonds, so all the bond angles at branch points would be 109.5°. Vulcanization involves branching with carbon-sulfur single bonds, so the bond angle would still be 109.5°.

CHAPTER NINETEEN

Transition Metals

19.1 Ca: [Ar] $4s^2$
Sc: [Ar] $4s^2 3d^1$

19.3 Ni^{2+} is [Ar] $3d^8$
Cu^+ is [Ar] $3d^{10}$
Zn^{2+} is [Ar] $3d^{10}$
Ti^{2+} is [Ar] $3d^2$

19.5 $[Cr(H_2O)_4Cl_2]Cl$ forms two ions, $[Cr(H_2O)_4Cl_2]^+$ and Cl^-. Only one chloride ion can be precipitated; the other two chlorides are ligands in the chromium complex ion.

19.7 $Pd(NH_3)_2Cl_2$ has zero free chloride ions and zero total ions.
$[Pd(NH_3)_3Cl]Cl$ has one free chloride ion and two total ions.
$Pd(NH_3)_4]Cl_2$ has two free chloride ions and three total ions.

19.9 Two isomers are available to this octahedral structure:

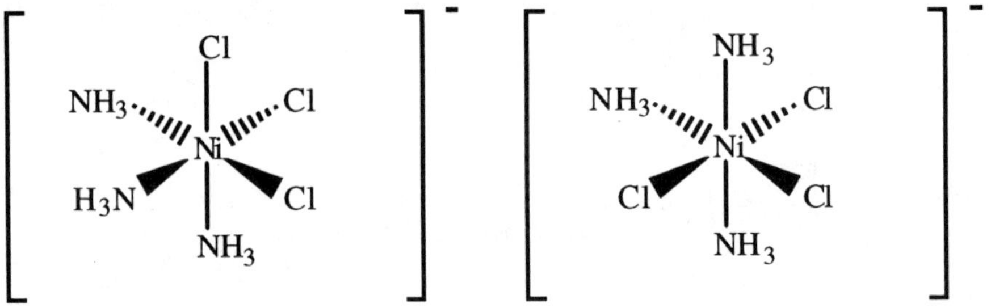

[structures of two octahedral Ni complex isomers with Cl and NH₃ ligands]

19.11 a. potassium tetracyanonickelate (II)
b. sodium trioxalatochromate(III)
c. pentaamminechloroplatinum(IV) chloride
d. tetraamminedinitroiron(III) sulfate
e. hexaaquacobalt(III) iodide
f. pentaamminechlororuthenium(III) bromide
g. potassium tetrachlorocobaltate (II)

19.13 a. For $CuCl_4^{2-}$ coordination number = **4**, oxidation number = **+2**.
b. For $Ag(CN)_2^-$ coordination number = **2**, oxidation number = **+1**.
c. For $Zn(NH_3)_4^{2+}$ coordination number = **4**, oxidation number = **+2**.

CHAPTER NINETEEN

19.15 The electron configuration of titanium is $[Ar]4s^23d^2$, so the Ti^{3+} ion is $[Ar]3d^1$. Therefore, Ti^{3+} has only one unpaired electron, regardless of the geometry of the crystal field.

19.17 Mn^{2+} has five d-electrons; in an octahedral crystal field, there could be either **five** or **one** unpaired electron(s).
Co^{3+} has six d-electrons; in an octahedral crystal field, there could be either **zero** or **four** unpaired electrons.
Ni^{2+} has eight d-electrons; in an octahedral crystal field, there could only be **two** unpaired electrons.

19.19 Fe^{2+} has six d-electrons; in an octahedral crystal field, there could be either zero or four unpaired electrons. If the complex is paramagnetic, there must be four unpaired electrons. The complex is high-spin (low value of Δ_o).

19.21. The energy of the transition can be calculated from the equation $E = hc/\lambda$,

$E = (6.626 \times 10^{-34} \text{ J·s})(3.00 \times 10^8 \text{ m/s})/(6 \times 10^{-7} \text{ m})$

$= 3.3 \times 10^{-19}$ J

$\Delta_o = E = 3.3 \times 10^{-19}$ J

Since 600 nm light is orange, the solution should appear to be blue, the complementary color of orange.

19.23 a. Cr^{3+} contains 3 d-electrons. It is **paramagnetic**.
b. Ni^{2+} contains 8 d-electrons. It is **paramagnetic**.
c. Co^{2+} contains 7 d-electrons. It is **paramagnetic**.
d. Fe^{2+} contains 6 d-electrons. It can be either **paramagnetic** (high spin) or **diamagnetic** (low spin).
e. Fe^{3+} contains 5 d-electrons. It is **paramagnetic**.
f. Cu^{2+} contains 9 d-electrons. It is **paramagnetic**.

19.25 Octahedral coordination requires 6 water molecules:
$Cr^{3+} + 6H_2O(l) \rightarrow [Cr(H_2O)_6]^{3+}(aq)$

19.27 $CrCl_3$ is acidic in water due to loss of protons from the aquo complex:
$CrCl_3 + 6H_2O \rightarrow Cr(H_2O)_6^{3+} + 3Cl^-$, K = very large
$Cr(H_2O)_6^{3+} + H_2O \rightarrow Cr(H_2O)_5(OH)_2 + H_3O^+$, $K_a = 1.26 \times 10^{-4}$
(10.0 g $CrCl_3$) (1 mol/158.36 g) = 0.0631 mol $CrCl_3$
0.0631 mol $CrCl_3$ will produce 0.0631 mol $Cr(H_2O)_6^{3+}$.

CHAPTER NINETEEN

If we let "x" moles per liter of $Cr(H_2O)_6^{3+}$ dissociate, then
$[Cr(H_2O)_6^{3+}] = (0.0631 - x)$ and $[H_3O^+] = [Cr(H_2O)_5(OH)^{2+}] = x$.
$1.26 \times 10^{-4} = [H_3O^+][Cr(H_2O)_5(OH)^{2+}]/[Cr(H_2O)_6^{3+}]$
$1.26 \times 10^{-4} = x^2/(0.0631 - x)$

We will assume that $(0.0631 - x) \approx 0.0631$.

$x^2 \approx (1.26 \times 10^{-4})(0.0631)$; $x = 0.00282$

The assumption was only fair, so we will continue to solve by iteration:
$(0.0631 - x) \approx (0.0631 - 0.00282) = 0.0603$

$x \approx 0.00276$

$(0.0631 - x) \approx (0.0631 - 0.00276) = 0.0603$

$x \approx 0.00276$

$[H_3O^+] = 2.76 \times 10^{-3}$ M so **pH = 2.559**

19.29 $Co^{3+} + 6NH_3 \rightarrow Co(NH_3)_6^{3+}$, $K = 5.0 \times 10^{31}$
$5.0 \times 10^{31} = [Co(NH_3)_6^{3+}]/[Co^{3+}][NH_3]^6 = (0.100)/[Co^{3+}](0.120)^6$
$[Co^{3+}] = 6.7 \times 10^{-28}$ M

19.31 a. Hexaaquacopper(II) chloride
b. Tetraamminedichloronickel(II)
c. Sodium tetrachloroplatinate(II)
d. Potassium hexacyanoferrate(II)

19.33 The density of the magnetite must be the same as the density of its unit cell. The volume of the cubic unit cell is equal to $(8.37 \text{ Å})^3 = (8.37 \times 10^{-8} \text{ cm})^3$
$= 5.86 \times 10^{-22}$ cm^3. We can calculate the mass of the unit cell from its volume and density:
  m = dV = $(5.18 \text{ g/cm}^3)(5.86 \times 10^{-22} \text{ cm}^3) = 3.04 \times 10^{-21}$ g
  $(3.04 \times 10^{-21} \text{ g})(6.022 \times 10^{23}/\text{mol}) = 1830$ g/mol

The molar mass of Fe_3O_4 is 232 g/mol, so the total number of formula units per unit cell is (1830 g/mol)/(232 g/mol) = 8 formula units per unit cell.
Iron has an average oxidation state of +8/3 in Fe_3O_4. Since individual atoms can have only integral oxidation states, this corresponds to two atoms in the +3 oxidation state and one atom in the +2 oxidation state. If there are eight formula units per unit cell, then there are sixteen Fe^{3+} and eight Fe^{2+} per unit cell.

CHAPTER TWENTY

Metallurgy

20.1 $Hg + 1/2\ O_2 \rightarrow HgO$

$\Delta H° = \Delta H_f°(HgO) = -90.83$ kJ/mol

$\Delta S° = \Sigma S°_{products} - \Sigma S°_{reactants}$
     $= 70.29$ J/mol·K $- [76.0$ J/mol·K $+ 1/2(205$ J/mol·K$)] = -108.2$ J/mol·K

$\Delta G° = \Delta H° - T\Delta S° = -90.83$ kJ/mol $- (298K)(-0.1082$ kJ/mol·K$) =$ **−58.6 kJ/mol**

20.3 The reaction is $CuO \rightarrow Cu + 1/2O_2$

From Table 12.4, $\Delta H_f°(CuO) = -155$ kJ/mol; ∴ ΔH of this reaction is $+155$ kJ/mol

$\Delta S° = 33.3$ J/mol·K $+ 1/2(205$ J/mol·K$) - 45.3$ J/mol·K $= 90.5$ J/mol·K

At the minimum necessary temperature, $\Delta G° = 0 = \Delta H° - T\Delta S°$

∴ $T = \Delta H°/\Delta S° = 155$ kJ/mol$/0.0905$ kJ/mol·K $=$ **1710K**

20.5 For the reaction $Fe_2O_3 + 2Al \rightarrow 2Fe + Al_2O_3$

$\Delta H° = -1676$ kJ/mol $- (-822$ kJ/mol$) = -854$ kJ

$\Delta S° = 50.92$ J/mol·K $+ 2(27.2$ J/mol·K$)-90.0$ J/mol·K $- 2(28.33$ J/mol·K$) = -41.34$ J/K

$\Delta G° = -854,000$ J $- 1273K(-40.21$ J/·K$) = -801,400$ J

$\ln K = -\Delta G°/RT = -(-801,400$ J$)/[(8.314$ J/·K$)(1273K)] = 75.71$

K = 7.6 × 10³²

20.7 For the reaction $PbO + C \rightarrow CO + Pb$

$\Delta H° = -110$ kJ/mol $- (-217$ kJ/mol$) = 107$ kJ/mol

$\Delta S° = 198$ J/mol·K $+ 65$ J/mol·K $- 68$ J/mol·K $- 5.7$ J/mol·K $= 189$ J/mol·K

$\Delta G° = \Delta H° - T\Delta S° = 0$ at the minimum temperature.

$T = \Delta H°/\Delta S° = (107000$ J/mol$)/(189$ J/mol·K$) =$ **566K**

20.9 From Problem 20.7, we have $\Delta H° = 107$ kJ/mol, $\Delta S° = 189$ J/mol·K, and so

$\Delta G° = 107000$ J/mol $- (1000K)(189$ J/mol·K$) = -82000$ J/mol

$\ln K = -\Delta G°/RT = -(-82000$ J/mol$)/[8.314$ J/mol·K $(1000K)] = 9.86$

K = 19000

20.11 $CuO + H_2 \rightarrow Cu + H_2O$

$\Delta H° = -242$ kJ/mol $- (-155$ kJ/mol$) = -87$ kJ/mol

$\Delta S° = 33.3$ J/mol·K $+ 189$ J/mol·K $- 43.5$ J/mol·K $- 130.6$ J/mol·K $= 48.2$ J/mol·K

$\Delta G°_{500°C} = -87000$ J/mol $- (500 + 273.15)K(4.82$ J/mol·K$) =$ **−124 kJ/mol**

$\Delta G°_{1000°C} = -87000$ J/mol $- (1000 + 273.15)K(48.2$ J/mol·K$) =$ **−148 kJ/mol**

CHAPTER TWENTY

20.13 For this problem, calculate $\Delta H°$ and $\Delta S°$ from Tables of $\Delta H°_f$ and $S°$, respectively. Then use $\Delta G° = \Delta H° - T\Delta S°$ assuming that $\Delta H°$ and $\Delta S°$ are independent of temperature. Finally, calculate ln K at five temperatures using $\ln K = -\Delta G°/RT$.

(1) For $CaO + H_2 \rightarrow Ca + H_2O$

$\Delta H° = -155$ kJ/mol $- (-635.09$ kJ/mol$) = 480.09$ kJ/mol

$\Delta S° = 41.6 + 70 - 130.6 - 40 = -58.4$ J/mol-K

$\Delta G° = 480{,}090 - T(-58.4) = -(8.314T) \ln K$.

Now calculate $\Delta G°$ and K at five temperatures. Here is a typical set of numbers:

T, K	$\Delta G°$, kJ	K
500	509.3	6.2×10^{-54}
1000	538.5	7.4×10^{-29}
1500	567.7	1.7×10^{-20}
2000	596.9	2.6×10^{-16}
3000	655.3	3.9×10^{-12}

A graph of K vs. T would show that K is always much less than 1. That is, the reaction is never spontaneous under standard conditions.

(2) For $CaO + C \rightarrow Ca + CO$

$\Delta H° = -110$ kJ/mol $- (-635.09$ kJ/mol$) = 525.09$ kJ/mol

$\Delta S° = 41.6$ J/mol·K $+ 198$ J/mol·K $- 5.7$ J/mol·K $- 40$ J/mol·K $= 193.9$ J/mol·K

$\Delta G° = 525{,}090 - T(193.9) = -(8.314T) \ln K$

Here are the results for the same five temperatures:

T, K	$\Delta G°$, kJ	K
500	428.1	1.9×10^{-45}
1000	331.2	5.0×10^{-18}
1500	234.2	7.0×10^{-9}
2000	137.3	2.6×10^{-4}
3000	-56.6	9.7

Now a graph of K vs T shows K becoming larger than 1. $\Delta G°$ passes through zero at about 2700K; above this temperature, the reaction is spontaneous.

20.15 a. $2Fe_2O_3 + 3C \rightarrow 4Fe + 3CO_2$

$Fe_2O_3 + 2Al \rightarrow 2Fe + Al_2O_3$

b. $CuO + H_2 \rightarrow Cu + H_2O$

$3CuO + 2Al \rightarrow 3Cu + Al_2O_3$

c. $Ag_2S + 2Na \rightarrow Na_2S + 2Ag$

$Ag_2S + Ca \rightarrow CaS + 2Ag$

CHAPTER TWENTY

20.17 $Sb_2S_3 + 2Fe \rightarrow Fe_2S_3 + 2Sb$

20.19 Each atom of sodium metal deposited will require 1 electron to reduce it, so the equation would be (25.0 A)(1 hr) (3600 s/hr)(23.00 g/mol)/(96500 C/mol) = **21.4 g**

20.21 We write the reaction as
MO + C → M + CO $\Delta H°$ and $\Delta S°$ for any such reaction can be obtained from standard enthalpies of formation and entropy values:
$\Delta H° = \Delta H°_{f\,CO} - \Delta H°_{f\,MO}$
$\Delta S° = S_M + S_{CO} - S_{MO} - S_C$
Then $\Delta G° = \Delta H° - T\Delta S°$; at the minimum temp for reduction, $\Delta G° = 0$, so
$T = \Delta H°/\Delta S°$ where we assume that $\Delta H°$ and $\Delta S°$ do not vary with temperature.
Substituting, we get
$T = [\Delta H°_{f\,CO} - \Delta H°_{f\,MO})]/[S°_M + S°_{CO} - S°_{MO} - S°_C]$.

20.23 a. $TiO_2(s) + 2Cl_2(g) \rightarrow TiCl_4(l) + CO_2(g)$
b. $2\,Al_2O_3(s) + 3C(s) \rightarrow 4Al(s) + 3CO_2(g)$

20.25 Tungsten is typically found in nature in the +6 oxidation state, as in $FeWO_4$.
1.00 kg of W = $(1.00 \times 10^3$ g$)/183.85$ g/mol = 5.44 moles of W
Since one mole of H_2 has two moles of reducing capacity, we need +6/2 = 3 moles of H_2 per mole of W.
5.44 moles × 3 = 16.3 moles of H_2 needed
We can calculate the volume the hydrogen gas would occupy at 800°C from V = nRT/P:
V = (16.3 mol)(0.08206 L·atm/mol·K)(1073.15K)/1.00 atm = **1440 L**

CHAPTER TWENTY-ONE

Organic Chemistry

21.1 There are five isomeric hexanes. Their structures and IUPAC names are:

$$CH_3(CH_2)_4CH_3 \quad \text{n-hexane}$$

$$CH_3(CH_2)_2-\underset{\underset{CH_3}{|}}{CH}-CH_3 \quad \text{2-methyl pentane}$$

$$CH_3-CH_2-\underset{\underset{CH_3}{|}}{CH}-CH_2CH_3 \quad \text{3-methyl pentane}$$

$$CH_3CH_2-\underset{\underset{CH_3}{|}}{\overset{\overset{CH_3}{|}}{C}}-CH_3 \quad \text{2,2-dimethyl butane}$$

$$\underset{\underset{CH_3}{|}}{CH_3CH}-\underset{\underset{CH_3}{|}}{CH}-CH_3 \quad \text{2,3-dimethyl butane}$$

21.3 a. The structures of the nine possible isomeric heptanes are given in Table 21.5.
 b. Their IUPAC names are n-heptane; 2-methylhexane; 3-methylhexane; 2,2-dimethylpentane; 2,3-dimethylpentane; 2,4-dimethylpentane; 3,3dimethylpentane; 3-ethylpentane; and 2,2,3-trimethylbutane.
 c. 2-methylhexane might commonly be called *iso*heptane because it has the simplest branching pattern.
 d. 2,2,3-trimethylbutane might commonly be called *neo*heptane because it is the isomer with the greatest amount of branching.
 e. The names used in (c) and (d) are *not* unambiguous.

21.5 a. $CH_3CH_2CH_3$ + heat (750°C) → $CH_2=CHCH_3$ + H_2
 $CH_2=CHCH_3 + H_2O$ (in dilute acid) → $CH_3\text{-}CH(OH)\text{-}CH_3$
 b. $CH_3\text{-}CH(OH)\text{-}CH_3 + H_2SO_4$ (180°C) → $CH_3CH=CH_2 + H_2O$
 $CH_3CH=CH_2 + H_2$ (and catalyst) → $CH_3CH_2CH_3$

21.7 Acetylene is C_2H_2 and has the Lewis structure H-C≡C-H
 If a proton is removed, we have the acetylide anion, [H-C≡C:]$^-$
 Similarly, the dianion is [:C≡C:]$^{2-}$

CHAPTER TWENTY-ONE

21.9 There are 12 possible methylbenzenes:

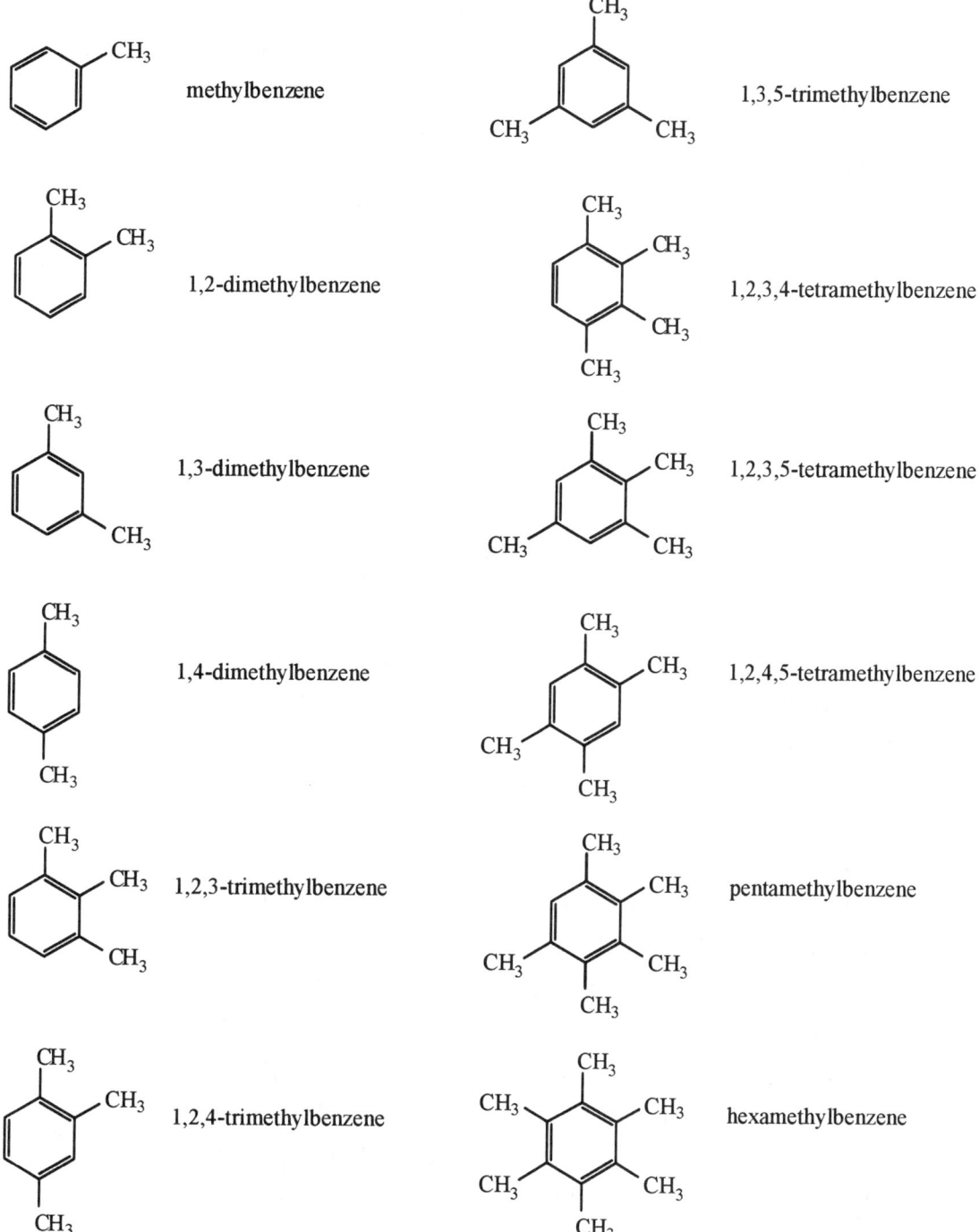

CHAPTER TWENTY-ONE

21.11 In general, the smaller the hydrocarbon portion of the molecule, the more "water-like" the alcohol will be. Ethanol is freely miscible in water. Phenol ionizes more than aliphatic alcohols, owing to the stability of the phenoxide ion. 1,6-dihydroxyhexane is a diol and is more soluble than alcohols of the same molar mass. 2-phenylethanol has the largest hydrocarbon chain of all these alcohols and does not ionize as phenol does. It has the lowest solubility of this group.
Solubility in water:
$CH_3CH_2OH > C_6H_5OH > HO(CH_2)_6OH > CH_3(CH_2)_5OH > C_6H_5(CH_2)_2OH$

21.13 a. There are four monobromobutanes:

$CH_3CH_2CH_2CH_2Br$ 1-bromobutane

$CH_3-CH(CH_3)-CH_2Br$ 2-bromobutane

$CH_3CH_2-CH(Br)-CH_3$ 1-bromo-2-methylpropane

$(CH_3)_3C-Br$ 2-bromo-2-methylpropane

b. $CH_3(CH_2)_3Br + OH^- \rightarrow CH_3(CH_2)_3OH + Br^-$
$(CH_3)_2CHCH_2Br + OH^- \rightarrow (CH_3)_2CHCH_2OH + Br^-$
$(CH_3CH_2-CH(Br)-CH_3) + OH^- \rightarrow CH_3CH_2CH(OH)CH_3 + Br^-$
$(CH_3)_3CBr + OH^- \rightarrow (CH_3)_3COH + Br^-$

c. i. $CH_3(CH_2)_3OH + O_2 \rightarrow CH_3(CH_2)_2COOH + H_2O$
    $(CH_3)_2CHCH_2OH + O_2 \rightarrow (CH_3)_2CHCOOH + H_2O$
ii. $2 (CH_3)_2CHCH_2OH + O_2 \rightarrow 2(CH_3)_2CHCHO + 2H_2O$
    $2 CH_3(CH_2)_3OH + O_2 \rightarrow 2CH_3(CH_2)CHO + 2H_2O$
iii. $CH_3CH_2CH(OH)CH_3 + O_2 \rightarrow 2 CH_3CH_2COCH_3 + 2H_2O$
iv. $(CH_3)_3COH$ cannot be further oxidized without disrupting the carbon skeleton.

d. $CH_3(CH_2)_3Br$ + alcoholic KOH $\rightarrow CH_3CH_2CH=CH_2$
$CH_3CH_2-CH(Br)-CH_3$ + alc. KOH $\rightarrow CH_3CH=CHCH_3$ or $CH_3CH_2CH=CH_2$
$(CH_3)_2CHCH_2Br$ + alc. KOH $\rightarrow (CH_3)_2C=CH_2$ (+KBr)
$(CH_3)_3C-Br$ + alc. KOH $\rightarrow (CH_3)_2C=CH_2$ (+KBr)

CHAPTER TWENTY-ONE

21.15 a.

      acetic acid   α-bromoacetic acid

b. HA → H$^+$ + A$^-$, K$_a$ = [A$^-$][H$^+$] / [HA] = x^2/(0.10 − x)
For acetic acid, pK$_a$ is 4.7, so K$_a$ = $10^{-4.7}$ = 2.01 × 10^{-5}.
2.01 × 10^{-5} = x^2/(0.10 − x)
x = 1.412 × 10^{-3} = [H$^+$], **pH = 2.85**
For bromoacetic acid, pK$_a$ is 2.5, so K$_a$ = $10^{-2.5}$ = 3.2 × 10^{-3}.
3.2 × 10^{-3} = x^2/(0.10 − x)
x = 1.6 × 10^{-2} = [H$^+$], **pH = 1.78**

c. The calculations in part b demonstrate that α-bromoacetic acid is more acidic than acetic acid. This is because the electronegative bromine atom in α-bromoacetic acid stabilizes the α-bromoacetate anion that forms as protons dissociate. In general, electronegative substituents tend to increase the acidity of carboxylic acids. This is called the inductive effect.

21.17 a. CH$_3$COOH + NH$_3$ + heat → CH$_3$CONH$_2$ + H$_2$O
CH$_3$COOH + CH$_3$OH → CH$_3$COOCH$_3$ + H$_2$O
CH$_3$COOH + H$_2$O → no reaction (other than to ionizes slightly)

b. CH$_3$COOCH$_3$ + NH$_3$ → no reaction
CH$_3$COOCH$_3$ + CH$_3$CH$_2$OH → no reaction
CH$_3$COOCH$_3$ + H$_2$O + OH$^-$ → CH$_3$COO$^-$ + CH$_3$OH + H$_2$O

c. CH$_3$COCl + NH$_3$ → CH$_3$CONH$_2$ + HCl (as NH$_4$Cl)
CH$_3$COCl + CH$_3$OH → CH$_3$COOCH$_3$ + HCl
CH$_3$COCl + H$_2$O → CH$_3$COOH + HCl

d. CH$_3$CONH$_2$ + CH$_3$COCl → CH$_3$CONHOCCH$_3$ + HCl
CH$_3$CONH$_2$ + CH$_3$OH → no reaction
CH$_3$CONH$_2$ + H$_2$O + H$^+$ → CH$_3$COOH + NH$_4$$^+$

CHAPTER TWENTY-ONE

21.19 There are four isomeric butylamines.

$CH_3CH_2CH_2CH_2NH_2$ 1-aminobutane

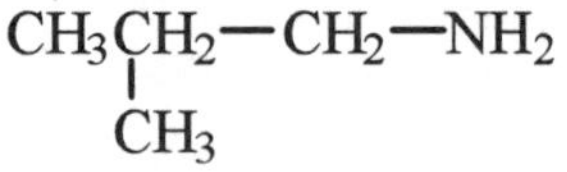

$CH_3CH_2-CH_2-NH_2$
$\quad\quad |$
$\quad\quad CH_3$ 2-aminobutane

$CH_3CH_2-CH-CH_3$
$\quad\quad\quad |$
$\quad\quad\quad NH_2$ 1-amino-2-methylpropane

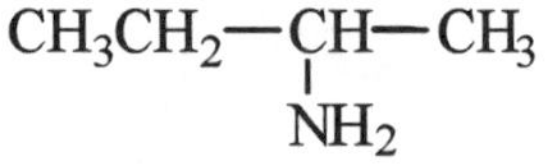

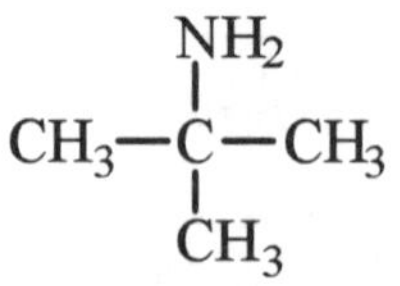

$\quad\quad NH_2$
$\quad\quad |$
CH_3-C-CH_3
$\quad\quad |$
$\quad\quad CH_3$ 2-amino-2-methylpropane

21.21 a. $CH_3CH_2CH_2OH + H_2SO_4(catalyst) \rightarrow CH_3CH=CH_2 + H_2O$
$\quad\quad CH_3CH=CH_2 + H_2O \rightarrow (CH_3)_2CHOH$
b. $(CH_3)_2CHCl + $ alcoholic $KOH \rightarrow CH_3CH=CH_2 + KCl + H_2O$
$\quad\quad CH_3CH=CH_2 + HBr + $ peroxides $\rightarrow CH_3CH_2CH_2Br$
c. $(CH_3)_2CHOH + H_2SO_4(catalyst) \rightarrow CH_3CH=CH_2 + H_2O$
$\quad\quad CH_3CH=CH_2 + HCl \rightarrow (CH_3)_2CHCl$
$\quad\quad (CH_3)_2CHCl + NH_3 \rightarrow (CH_3)_2CHNH_2 + HCl$
d. $(CH_3CO)_2O + H_2SO_4(catalyst) + CH_3CH_2OH \rightarrow CH_3CO_2CH_2CH_3$
e. $CH_3CH_2OH + O_2 \rightarrow CH_3COOH$

21.23 a. Aldehydes reduce Tollen's reagent so "A" must be formaldehyde.
$\quad\quad H_2CO(l) + 2Ag(NH_3)_2^+(aq) + 2H_2O \rightarrow HCO_2^-(aq) + Ag(s) + 4NH_4^+(aq) + OH^-(aq)$
b. "B" and "C" must be isopropanol and acetone, respectively.
$\quad\quad 5(CH_3)_2CHOH + 2MnO_4^-(aq) \rightarrow 5(CH_3)_2CO + 2Mn^{2+}(aq) + 6OH^-(aq) + 2H_2O$

CHAPTER TWENTY-ONE

21.25 $C_5H_{11}Cl$ corresponds to a monosubstituted compound of a saturated hydrocarbon having the formula C_nH_{2n+2}. This particular parent hydrocarbon is pentane.. The eight isomers of chloropentane and their IUPAC names are:

$CH_3-CH_2-CH_2-CH_2-CH_2-Cl$ 1-chloropentane

$CH_3-CH_2-CH_2-\underset{\underset{Cl}{|}}{CH}-CH_3$ 2-chloropentane

$CH_3-CH_2-\underset{\underset{Cl}{|}}{CH}-CH_2-CH_3$ 3-chloropentane

$CH_3-CH_2-\underset{\underset{CH_3}{|}}{CH}-CH_2-Cl$ 1-chloro-2-methylbutane

$CH_3-\underset{\underset{CH_3}{|}}{CH}-CH_2-CH_2-Cl$ 1-chloro-3-methylbutane

$CH_3-CH_2-\overset{\overset{CH_3}{|}}{\underset{\underset{Cl}{|}}{C}}-CH_3$ 2-chloro-2-methylbutane

$CH_3-\underset{\underset{CH_3}{|}}{CH}-\underset{\underset{Cl}{|}}{CH}-CH_3$ 2-chloro-3-methylbutane

$CH_3-\overset{\overset{CH_3}{|}}{\underset{\underset{CH_3}{|}}{C}}-CH_2-Cl$ 1-chloro-2,2-dimethylpropane

21.27 Halogenation (e.g. bromination) and oxidation of propylene result in addition across the double bond.
$CH_3CH=CH_2 + Br_2 \rightarrow CH_3CH(Br)-CH_2Br$
$CH_3CH=CH_2 + KMnO_4 \rightarrow MnO_2 + CH_3CH(OH)-CH_2OH$
In contrast, halogenation and oxidation of toluene do *not* involve the double bond. Benzene rings are stabilized by the aromatic cyclic, conjugated pi-systems. It would take a lot of energy to disrupt such a stable system. Instead, aromatic chemistry is predominantly substitution chemistry or the chemistry of appended functional groups.
$C_6H_5CH_3 + Br_2 + Fe \rightarrow$ brominated toluenes (e.g. $C_6H_4BrCH_3$)
$C_6H_5CH_3 + KMnO_4 \rightarrow C_6H_5CO_2H$

CHAPTER TWENTY-ONE

21.29 Benzoic acid, like all carboxylic acids, is strongly hydrogen bonding. In acetone, it can hydrogen bond with the carbonyl groups on the acetone molecules. Carbon tetrachloride is nonpolar.

Even though benzoic acid is soluble in carbon tetrachloride, it cannot hydrogen bond with this solvent and instead forms dimer pairs by hydrogen bonding to itself:

$2C_6H_5COOH \rightarrow (C_6H_5COOH)_2$

This accounts for the doubling in apparent molar mass through boiling point elevation data: $2(122) = 244$.

21.31 Propylene is known to react with bromine by adding across the double bond:
$CH_3CH=CH_2 + Br_2 \rightarrow CH_3CHBrCH_2Br$

Propane is unreactive with bromine in the absence of light radiation. We can determine the quantity of propylene in the mixture from the amount of bromine that is required to add across the double bond.
$(24.7 \text{ g } Br_2)(1 \text{ mol}/159.81 \text{ g}) = 0.154 \text{ mol } Br_2$.

Since the reaction stoichiometry is 1:1, this also reacted with 0.154 mol propylene.
$(0.154 \text{ mol } CH_3CH=CH_2)(42.08 \text{ g}/1 \text{ mol}) = 6.48 \text{ g } CH_3CH=CH_2$

Thus, the original mixture contained 6.48 g propylene and 3.52 g propane. The percentage of propane in the mixture is **35.2%**.

21.33 The balanced equations for the process are:
$CO(g) + NaOH(aq) \rightarrow NaHCO_2(aq)$
$NaHCO_2(aq) + H_3O^+(aq) \rightarrow Na^+(aq) + H_2O(liq) + HCO_2H(aq)$
$HCO_2H(aq) + OH^-(aq) \rightarrow H_2O(liq) + HCO_2^-(aq)$

The theoretical yield for the formic acid can be calculated from the quantity of carbon monoxide that reacts.

$n_{CO} = PV/RT = (4.50 \text{ atm})(0.100 \text{ L})/(0.08206 \text{ L·atm/mol·K})(298K)$
   $= 0.0184 \text{ mol CO}$
$(0.0184 \text{ mol CO})(1 \text{ mol } HCO_2H/1 \text{ mol CO}) = 0.0184 \text{ mol } HCO_2H$
$(0.0184 \text{ mol } HCO_2H)(46.02 \text{ g/mol}) = 0.847 \text{ g } HCO_2H)$

The actual yield of formic acid can be calculated from the titration data.
   $(0.123 \text{ L OH}^-)(0.100 \text{ mol/L}) = 0.0123 \text{ mol OH}^-$
   $(0.0123 \text{ mol OH}^-)(1 \text{ mol } HCO_2H/\text{mol OH}^-) = 0.0123 \text{ mol } HCO_2H$
   $(0.0123 \text{ mol } HCO_2H)(46.02 \text{ g/mol}) = 0.566 \text{ g } HCO_2H$

Yield $= 0.566 \text{ g}/0.847 \text{ g} = 0.668 =$ **66.8%**

CHAPTER TWENTY-ONE

21.35 Mylar is an ester condensation polymer between dimethyl terephthalate and ethylene glycol monomer subunits. The formula for dimethyl terephthalate is $CH_3OOC(C_6H_4)COOCH_3$ (M = 194.19 g/mol), and the formula for ethylene glycol is $HO(CH)_2OH$ (M = 60.05 g/mol). One mole of methanol is produced for every mole of dimethyl terephthalate and ethylene glycol that condense; this also produces a polymer with a mass of (194.19 + 60.05 − 32.04) g = 222.19 g. $(1.00 \times 10^5$ g mylar)(1 mol CH_3OH/222.19 g mylar) = 450. mol CH_3OH.

We can now calculate the volume of methanol from its mass and density:
V = m/d = (450. mol)(32.04 g/mol)/(0.79 g/cm³)
V = 1.8×10^4 cm³ = **18 L**

21.39 The formula C_4H_8 is of the type C_nH_{2n}, suggesting either a ring or a single degree of unsaturation.

Since alkenes react with dilute aqueous permanganate to produce diols, the only structures with this empirical formula that would not react with dilute aqueous permanganate must be saturated hydrocarbons. The only two possibilities with the formula C_4H_8 are the four-membered ring cyclobutane, and methyl cyclopropane.

CHAPTER TWENTY-TWO

Nuclear Chemistry

22.1 a. The mass number, A, is indicated by the superscript next to the element's symbol. The mass number for this element is **10**.
b. The atomic number, Z, is indicated by the subscript next to the element's symbol. The atomic number for this element is **5**.
c. The proton number of an element is equal to its atomic number, Z. This element has **5 protons**.
d. Electrons are the only extranuclear particles in an atom. In a neutral atom, there are the same number of electrons as protons.
The number of electrons is **5**.
e. This nuclide has **5 protons** (Z = 5). The number of neutrons is equal to the mass number, A − Z. Therefore, it also contains 10 − 5 = **5 neutrons**.
f. $_5B^{10}$ has Z = 5 and A = 10; therefore, this element is **Boron-10**.

22.3 $_{11}Na^{22} \rightarrow {_1\beta^0} + {_{10}Ne^{22}}$ β⁺ emission to correct having excess protons

$_{11}Na^{24} \rightarrow {_{-1}\beta^0} + {^{22}Mg_{12}}$ β⁻ emission to correct having excess neutrons

22.5 $_{88}Ra^{226} \rightarrow {_2He^4} + {_{86}Rn^{222}}$

22.7 $_6C^{14} \rightarrow {_{-1}\beta^0} + {_7N^{14}}$

22.9 a. $_5B^{10} + {_2He^4} \rightarrow {_6C^{13}} + {_1H^1}$
b. $_4Be^9 + {_1H^1} \rightarrow {_4Be^8} + {_1H^2}$ or $_4Be^9 + {_1H^1} \rightarrow {_4Be^8} + {_1D^2}$
c. The correct starting isotope has to be Al-27:
$_{13}Al^{27} + {_0n^1} \rightarrow \gamma + {_{13}Al^{28}}$

22.11 Magnesum-23 undergoes decay by positron emission:
$_{12}Mg^{23} \rightarrow {_1\beta^0} + {_{11}Na^{23}}$

22.13 $_7N^{14} + {_2He^4} \rightarrow {_1p^1} + {_8O^{17}}$

22.15 After 4 half-lives, the amount of material that has not undergone radioactive decay is $(1/2)^4$ or **6.25%**

22.17 a. After two half-lives, the decay rate will have decreased to $(1/2)^2 = 0.25$ of its initial value.
Decay rate = 250. counts per second

CHAPTER TWENTY-TWO

b. The decay rate will be 30 counts per second when $N = 0.030\ N_o$.
ln (1/0.030) = kt = 0.693 $t/t_{1/2}$;
t = (2.4 min/0.693) ln (1/0.030) = 12 min
The decay rate will be 30. counts/s **12 minutes** after it was 1000. counts/s.

22.19 Since radioactive decay is first order, the decay rate = kN.
k = $(0.693/t_{1/2})$ = (0.693/1620 yr) = 4.28×10^{-4} yr^{-1}
N = (1 g)(1 mol/226 g)(6.022×10^{23} atoms/mol) = 2.66×10^{21} atoms
Decay rate = (4.28×10^{-4} yr^{-1})(2.66×10^{21}) = 1.14×10^{18} counts/year
= **3.61×10^{10} counts/s**

22.21 For 10. g of C, the decay rate would be 150 counts/min in a living sample.
ln N_o/N = ln R_o/R = kt = ln (150/100.) = (0.693/5720 yr)(t)
t = 3400 yr

22.23 E = energy released = $-\Delta mc^2$; c^2 is the speed of light squared. If we use 1 amu as the mass in this equation, we find the energy to be
(1.00×10^{-3} kg/6.022×10^{23})(3.00×10^8 m/s)^2 = 1.49×10^{-10} J
1.49×10^{-10} J/1.602×10^{-19} eV/J = 931×10^6 eV or 931 MeV

This is a useful unit conversion amount for mass to energy in problems involving nuclear binding energy.
$-\Delta m = 2m_p$ = 2(1.0073 amu) = 2.0146 amu
E = (2.0146 amu)(931 MeV/amu) = **1.88×10^3 MeV**

22.25 Binding energy = $-\Delta mc^2$

a. H^2 contains one proton and one neutron, and its mass is 2.0140 amu.
Δm = [2.0140 – 1.0073 – 1.0087] amu = –0.0020 amu
Binding energy = $-\Delta mc^2$ = (0.0020 amu)(931 MeV/amu)
= **1.9 MeV**
Binding energy per nucleon = (1.9 MeV)/(2 nucleons)
= **0.95 MeV/nucleon**

b. C^{12} contains six protons and six neutrons, and its mass is 12.0000 amu.
Δm = [12.0000 – 6(1.0073) – 6(1.0087)] amu = –0.0960 amu
Binding energy = $-\Delta mc^2$ = (0.0960 amu)(931 MeV/amu)
= **89.4 MeV**
Binding energy per nucleon = (89.4 MeV)/(12 nucleons)
= **7.45 MeV/nucleon.**

CHAPTER TWENTY-TWO

c. Fe^{56} contains 26 protons and 30 neutrons, and its mass is 55.9349 amu.

$\Delta m = [55.9349 - 26(1.0073) - 30(1.0087)]$ amu $= -0.5159$ amu

Binding energy $= -\Delta mc^2 = (0.5159$ amu$)(931$ MeV/amu$)$

$= $ **480.3 MeV**

Binding energy per nucleon $= (480.3$ MeV$)/(56$ nucleons$)$

Binding energy per nucleon $= $ **8.58 MeV/nucleon**

22.27   Mass defect $= \Delta m$ for nucleus formation

Mass defect $= -[15.9994 - 8(1.007276) - 8(1.008665)]$ amu

$= $ **0.12813 amu**

Binding energy $= (0.12813$ amu$)(931.5$ MeV/amu$) = 119$ MeV

Binding energy per nucleon $= (119$ MeV$)/(16$ nucleons$)$

$= $ **7.46 MeV/nucleon**

22.29 $\Delta E = \Delta mc^2$; $\Delta m = (4.00260 + 1.00866 - 2.0140 - 3.01605)$ amu

$= -0.0188$ amu

$\Delta E = (-0.0188$ amu$)(931$ MeV/amu$) = -17.5$ MeV

So **17.5 MeV of energy are released** during the deuterium/tritium reaction given.

22.31 $_{92}U^{235} \rightarrow {_{90}Th^{231}} \rightarrow {_{91}Pa^{231}} \rightarrow {_{89}Ac^{227}} \rightarrow {_{90}Th^{227}} \rightarrow {_{88}Ra^{223}} \rightarrow {_{86}Rn^{219}} \rightarrow$

$^{217}Po_{84} \rightarrow {_{82}Pb^{213}} \rightarrow {_{83}Bi^{213}} \rightarrow {_{84}Po^{213}} \rightarrow {_{82}Pb^{209}}$

22.33 $_{90}Th^{232} \rightarrow {_{90}Th^{233}} \rightarrow {_{90}Th^{234}}$

$_{91}Pa^{233} \rightarrow {_{91}Pa^{234}}$

$_{92}U^{232} \rightarrow {_{92}U^{233}} \rightarrow {_{92}U^{234}} \rightarrow {_{92}U^{235}} \rightarrow {_{92}U^{236}} \rightarrow {_{92}U^{237}}$

$_{93}Np^{237}$

22.35   From Figure 22.7, we see that a nucleus with mass number $A = 152$ has a binding energy of about 8.1 MeV/nucleon, for a total binding energy of 152×8.1 MeV/nucleon $= 1230$ MeV. Nuclei with $A = 70$ to 80 have a binding energy of about 8.4 MeV/nucleon. The total binding energy for the two daughter nuclei is $(70 + 80) \times 8.4$ MeV/nucleon $= 1260$ MeV. The difference between these two energies is about **30 MeV**, and is the energy that was released to the surroundings during the fission of the heavier atom to the daughter atoms.

CHAPTER TWENTY-TWO

22.37 $_{92}U^{235} \rightarrow {}_{38}Sr^{90} + {}_{54}Xe^{143} + 2\ {}_{0}n^{1}$

22.39 a. $_{53}I^{131} \rightarrow {}_{-1}\beta^{0} + {}_{54}Xe^{131}$
b. $k = \ln(2/193.2\ hr) = 3.588 \times 10^{-3}\ hr^{-1}$
$\ln(R/R°) = -kt = -(3.588 \times 10^{-3}\ hr^{-1})(24.00) = -0.08610$
$e^{-0.08610} = 0.9175 = R/R°$
$R = (0.9175)(1500.) = $ **1376 counts/min**

22.41 Mass defect = (0.00915 amu/nucleon)(120 nucleons) = 1.10 amu
The mass of a Sn^{120} atom is 1.10 amu lighter than the mass of 50 separated protons and electrons and 70 separated neutrons.
Mass Sn^{120} = [50(1.0073 amu) + 50(0.0005486 amu) + 70(1.0087 amu)] − 1.10 amu
Mass = ^{120}Sn = 119.90 amu

22.43 a. $_{39}Y^{90} \rightarrow {}_{-1}\beta^{0} + {}_{40}Zr^{90}$
$_{29}Cu^{66} \rightarrow {}_{-1}\beta^{0} + {}_{30}Zn^{66}$
b. $_{94}Pu^{236} \rightarrow {}_{2}He^{4} + {}_{92}U^{232}$
$_{91}Pa^{226} \rightarrow {}_{2}He^{4} + {}_{89}Ac^{222}$

22.45 Each β-particle emission causes the atomic number to increase by one unit and it leaves the mass number unchanged. Each α-particle emission causes the atomic number to decrease by two units and the mass number to decrease by four units. Starting with $_{89}Ac^{231}$, the given series of radioactive emissions will produce
$_{90}Th^{231} \rightarrow {}_{91}Pa^{231} \rightarrow {}_{89}Ac^{227} \rightarrow {}_{87}Fr^{223} \rightarrow {}_{85}At^{219} \rightarrow {}_{83}Bi^{215} \rightarrow {}_{84}Po^{215}$
$\rightarrow {}_{82}Pb^{211}$

22.47 a. Average atomic mass = (0.909)(20 amu) + (0.0030)(21 amu) + (0.088)(22 amu)
Average atomic mass = **20.18 amu,** which is in agreement with the periodic table.
b. Let x = fractional abundance of U^{235}; (1 − x) = fractional abundance of U^{238}
(x)(235.044 amu) + (1 − x)(238.051 amu) = avg atomic mass = 238.029 amu
(238.051 − 238.029) = (238.051 − 235.044)x; x = 0.0073
Naturally occurring uranium is **0.73% U^{235} and 99.27% U^{238}**.

22.49 One possibility is the sequential bombardment by an alpha-particle and a neutron:
$_{12}Mg^{24} + {}_{2}He^{4} \rightarrow {}_{14}Si^{28}$
$_{14}Si^{28} + {}_{0}n^{1} \rightarrow {}_{11}Na^{22} + {}_{3}Li^{7}$

CHAPTER TWENTY-TWO

22.51 1.0 kg U^{235} = (1.0 × 10^3 g/235 g/mol)(6.023 × 10^{23} atoms/mol)
= 2.6 × 10^{24} atoms/kg
(2.6 × 10^{24} atoms/kg)(3.2 × 10^{-11} J/fission) = **8.2 × 10^{13} J/kg**

22.53 A metric ton is equal to 1 × 10^6 g or 1 × 10^3 kg.
Converting the heat content of uranium in GJ/kg to units of GJ/ton:
(460 GJ/1 kg)(1 × 10^3 kg/1 ton) = 4.6 × 10^5 GJ/ton of uranium metal.
The energy content of uranium metal is **1.6 × 10^4 to 2.4 × 10^4** times that of coal, given the manner in which each material is used for heat production.